Precision Farming

Premjit Sharma is having more than fifteen years of teaching and research experience mainly on cropping systems and soil fertility. Throughout his career, he has been associated with a number of educational and research institutes. He has published a number of research articles in reputed journals. His areas of interest include agriculture, agricultural economics, agricultural business management, natural resource management and sustainable development.

He has been associated with a dozen of UG and PG Courses in the discipline of agriculture and is the chairman of advisory committee of M.Sc. and Ph.D. students. Apart from being an active and dedicated teacher, he has also been committed in research and running projects funded by the eminent bodies.

Precision Farming

Editor
Dr. Premjit Sharma

2025
Gene-Tech Books
New Delhi - 110 002

ISBN 978-81-89729-62-2

Published by : **GENE-TECH BOOKS**
4762-63/23, Ansari Road, Darya Ganj,
NEW DELHI - 110 002
Phone: 43003222
e-mail: genetechbooks@yahoo.co.in

PRINTED IN INDIA

Preface

Precision farming is an emerging, important hi-tech approach to agricultural practices which promises to revolutionise the sector of agriculture. This new technology enables farmers to create finely detained maps of forms that describe important characteristics, such as fertiliser requirements, by a specific location of the field. Computer-aided machinery can thus assist the farmer in managing his fields better while responding to the spatial needs of his cropped areas.

Conventional approaches to farming respond to the agricultural lands' needs uniformly. Not only does precision farming rectify this lack of specificity, it also curbs excessive waste in the form of more efficient application of chemical inputs and decreased runoff of excess fertiliser into the environment.

This book not only introduces the reader to the technology of precision farming, it presents, a broad overview of its concepts and the tools of this systems. It also closely studies the mobility of this option and considers economic, environmental and other considerations. With a comprehensive insight into the subject, the book should prove to be an interesting read to the reader.

Editor

Contents

1

Introduction

Precision farming or precision agriculture is an agricultural concept relying on the existence of *in-field variability*. It requires the use of new technologies, such as global positioning (GPS), sensors, satellites or aerial images, and information management tools (GIS) to assess and understand variations. Collected information may be used to more precisely evaluate optimum sowing density, estimate fertilisers and other inputs needs, and to more accurately predict crop yields. It seeks to avoid applying same practices to a crop, regardless of local soil/climate conditions and may help to better assess local situations of disease or lodging.

In the Midwest it is associated not with sustainable agriculture but with mainstream farmers who are trying to maximise profits by spending.money only in areas that need fertiliser. This practice allows the farmer to vary the rate of fertiliser across the field according to the need identified by GPS guided Grid Sampling. Fertiliser that would have been spread in areas that don't need it can be placed in areas that do, thereby optimising its use.

Precision farming may be used to improve a field or a farm management from several perspectives :

— *agronomical perspective*: adjustment of cultural practices to take into account the real needs of the crop (e.g., better fertilisation management)

— *technical perspective*: better time management at the farm level (e.g. planification of agricultural activity)

— *environmental perspective*: reduction of agricultural impacts (better estimation of crop nitrogen needs implying limitation of nitrogen run-off)

— *economical perspective*: increase of the output and/or reduction of the input, increase of efficiency (e.g., lower cost of nitrogen fertilisation practice)

Other benefits for the farmer may be to help him set an history of his/her farm practices and results, to help him in his decision making and traceability requirements (as increasingly required in developed countries).

In addition to the assigned reading, consider the following statement in order to appreciate the challenge in reaching agreement on the name for a field that is emerging as rapidly as is precision farming. The statement comes from the Second International Conference on Site-Specific Management for Agricultural Systems, held in March 1994 in Minneapolis, Minnesota.

Site-specific crop management (SSCM) refers to a developing agricultural management system that promotes variable management practices within a field according to site or soil conditions. While this technology is only a few years old, various names have been used to describe the concept: farming by soil; farming soil, not fields; farming by the foot; spatially prescriptive farming; computer aided farming; farming by computer; farming by satellite; high-tech sustainable agriculture; soil-specific crop management; site-specific farming; and precision farming. Interest in this emerging concept has been featured in a variety of news media.

Site-specific crop management is an information and technology based agricultural management system to identify, analyse, and manage site-soil spatial and temporal variability within fields for optimum profitability, sustainability, and protection of the environment. SSCM employs a system engineering approach to crop production where inputs are made on an "as needed basis," and was made possible by recent innovation in information and technology such as microcomputers, geographic information systems, positioning technologies (Global Positioning System), and automatic control of farm machinery. It is a holistic approach to micro manage spatial and temporal variability in agricultural landscapes based on integrated soil, plant, information, and engineering management technologies as well as economies.

Yield Variability, Monitoring and Mapping

Crop yield relates to site-specific crop management, including the nature of yield variability, how it is measured, and how such measurements may be used to aid decision-making by growers and agricultural service producers.

Yield Variability

Precision agriculture requires integration of three elements: 1) positioning capabilities to know where equipment is located; 2) real-time mechanisms for controlling nutrient, pesticide, seed, water, or other crop production inputs; and 3) databases or sensors that provide information needed to develop input response to site-specific conditions. The technologies associated with requirements 1 and 2 are advanced compared with the understanding necessary to meet requirement 3. Building databases to quantify yield variability will improve the understanding of how various stresses affect plant growth, development, or yield, and ultimately lead to optimum site-specific prescriptions.

There is evidence that variation in nutrient elements in the soil is not the foremost factor affecting crop yields. Precision farming advocates have devoted significant efforts to applying fertilisers more selectively to soils based on yield potential and soil test results. The goal is to obtain more efficient use of applied fertiliser, to reduce any excess application that might cause environmental insult, and to improve economics. It should be emphasised that soil fertility needs, particularly nitrogen, are dependent on yield level and/or the amount of rainfall that occurs during the growing season. Variable rate application of fertiliser is based on expected yield, a parameter that is often difficult to predict. In northern Italy, corn yield variability could be explained by soil NO_3 content at the beginning of the growing season, and by spatial differences in soil carbon and nitrogen content.

Kitchen *et al* in Missouri found that variable-rate N fertiliser application resulted in lower residual soil nitrate, but showed little economic benefit. Ferguson *et al* in Nebraska detected no significant difference in corn grain yield or use efficiency between variable rate and single rate fertiliser application. Logsdon *et al* observed that crop yield variability was influenced by stored soil water in rain-fed agriculture. In a drier year soil water storage correlated with both corn and soybean yields. In a wet year, predicted aeration stress correlated with corn yields but not soybean yields.

Runge and Hons considered crop yield variations associated with both temporal and spatial factors that they had either experienced or obtained from the literature. Their intent was to focus sources of yield variation into three primary categories: water, fertility-management and genetics. They concluded that plant available stored soil water and seasonal precipitation quantity and distribution had the greatest effect on rainfed crop yields. To have a successful crop, producers know that a uniform stand at the optimum population density, adequate soil fertility and weed control are important. Farmers characteristically refer to areas of good and poor soils that are most often related to the variation in plant available stored water.

In a 16-ha field in Iowa, Karlen *et al* measured spatial variability in corn emergence, growth, and yield as well as in insect pressure and in several soil characteristics. Analysis of these data led them to conclude that in regard to developing site-specific management strategies, attention to soil water factors, soil acidity, and erosion prevention are more important than efforts to differentially apply fertiliser nutrients.

Variation in soil fertility and hydrologic properties across landscapes affects crop yield. Nolan et al showed that a simple landscape classification delimited areas within a rolling field of canola that responded differently to N fertiliser and that varying N fertiliser according to these landscape classes can increase profits from $5 to $7/ha compared with uniform application of N. Five landscape classes were recognised-upper level, shoulderslope, backslope, footslope, and lower level. Similarly, Khakural et al reported that corn and soybean yields were less at eroded slopes that an nearly level summits or at the foot/toeslope positions. Corn yield was positively correlated with A horizon thickness and negatively correlated with surface pH. A horizon thickness, surface pH, tillage system, and growing season precipitation explained 72% of the variability in corn yield.

Monitoring and Mapping

Early efforts to develop instantaneous yield monitors were concentrated on combine-harvested crops, perhaps because the relatively steady flow of grain into the harvester seemed to lend itself readily to measurement. Typical components of grain yield monitoring systems include grain flow sensors, grain moisture sensors, ground speed sensors, and a computer/display console. Campbell has reviewed the recent advances in yield monitoring of conveyor-harvested crops. In general, as explained by Hall et al, instantaneous yield for such crops is calculated by the equation

$$\text{Yield} = \frac{14.85\ \text{FR}}{\text{GS} \times \text{W}}$$

where:

Yield = crop yield (tons per acre)

FR = flow rate (pounds per second)

GS = ground speed (miles per hour)

W = harvester width (feet)

The constant in the numerator is specific for the units specified for the variables, and would, of course, be different if other units were used, for example, pounds per acre for flow rate, or if metric units are used for all variables. Essentially the same equation is suitable for computing instantaneous yields of combine-harvested crops, for which flow rate might be expressed in units of either mass or volume per unit time.

These pieces of information are gathered by yield monitors consisting of four basic components-product flow sensors, speed sensors, signal conditioning and conversion units, and a computer. The information is spatially referred with position data provided by a differentially corrected Global Positioning System (DGPS).

Yield monitors continue to be improved for use during the mechanised harvest of many crops, including conveyor-harvested crops, such as potatoes, tomatoes, and sugar beets, peanut, cotton, rice and other combine-harvested crops, including wheat, corn, and soybeans.

Yield monitoring and mapping is also being explored for crops that are harvested by hand labor rather than mechanised harvesters. For example, the yield of various fruits that are hand picked might be monitored by recording weight increases for picking boxes or bins and using DPGS to spatially reference each recorded weight. Perhaps this could be facilitated by using bins identified by bar codes and a forklift equipped with scales on its forks, DGPS receiver, and a bar code scanner. This would also facilitate tracking of bins of fruit from specific point of harvest as they proceed through the packing house.

Plant Nutrients, Soil Fertility and Grid Sampling

Early American settlers were first introduced to Precision Farming by friendly indians who taught them corn production and soil fertility practices such as placing a fish (or two) under each hill of corn to supply a slow release nutrient source of N-P-K and trace elements. The rates were no doubt based on visual assessments of soil depth and organic matter and past performance of the field.

As mechanical tillage machinery and methods revolutionised agriculture in the 19th and 20th centuries, the field became the major unit of production. Justus von Liebig's theories of the chemical basis of plant growth coupled with his now famous "Law of the Minimum" led the way for chemical analysis of soil. Total chemical analysis of soil left much to be desired from a practical viewpoint because early tests (and some today) bore no relationship to plant growth response since they did not measure plant availability or uptake of the nutrient ions by the crop.

By the 1930s and 40s many state experiment stations were establishing soil testing laboratories and urging farmers to assay their soil nutrient status. Most recommendations went to great lengths to assure that the sampling technique produced a random representative sample of the field. Many recommendations suggested one sample per 20 acres but this was often compromised to accommodate larger field sizes.

Following the end of World War II, fertiliser, especially nitrogen, became more generally and economically available. Agronomists and fertiliser sales reps brought Liebig's Law of the Minimum back into the philosophy of soil nutrient management, but by this time the list of known essential elements had expanded to sixteen.

Many agronomists and the Potash Institute expanded the idea of the limiting element, often portrayed as a barrel with one short stave or as a chain with a weakened link, to all management factors. Many researchers decried the approach as non-scientific or too muddled by interactions difficult to measure or assess. Two approaches to soil nutrient status were practiced. One attempted to build soil fertility levels by application of organic matter and amendments so that crops would not experience nutrient limitations.

The second was to supply the nutrients needed irregardless of the soil nutrient status. Reduced profit margins subsequently brought crop production managers back to a more practical approach of supplying only those elements that limited yield. Environmental concerns also made blanket applications of nutrients, pesticides or water politically, socially and biologically incorrect.

In the 1990s the availability of GPS and GIS to agricultural applications suddenly made it possible to manage very small units rather than managing the field as an average. Reawakened interest in the Law of the Minimum and soil testing held out the prospect of increasing yield while decreasing costs. In many instances soil differences and yield monitors justified the expense of the GPS, GIS technology as increased yields were achieved with fewer chemical inputs. Additionally, the environmental cost of over-application was lessened especially in the case of nitrate and phosphorus fertilisation.

New rules were developed for frequency and location of soil sampling to demonstrate the potential for cost savings. Many early Precision-ag practitioners advised sampling on a grid with each square representing 4 acres or 2 acres or <=1 acre. The grid pattern was quickly refined by practice and modified into diamond, random or other grid sampling schemes. Efforts to reduce costs and careful analysis of the data has had the effect of looking at combinations of sampling or soil zones based on soil type, yield maps, irrigation, drainage, salinity or other relevant parameters. This is the point at which precision ag now finds itself.

In the future scientists envision having probes mounted on tillage equipment which will map the entire fields for soil texture, pH, salinity or chemical parameters. The Veris technology represents one of the early ventures into this type of soil mapping. Very often the new map becomes the basis for more complete soil testing by zone.

Soil testing methods remain the subject of research and controversy. Recent discussions in California indicate that on Western soils, physical properties, salinity and compaction may be as important to yield as the chemical parameters which have played such a prominent role in the adoption of precision agricultural technology in the Midwest. Bringing the results of soil tests and yield monitoring together via map stacking and other data analysis tools (SST) has demonstrated that every field is not likely to present easy "textbook" answers.

Dr. Pierre Robert and others have emphasised that agronomic sleuthing or artistry and experience may become the most valuable tool required to take advantage of the precision ag technology. In the not too distant future, however, precision agriculture and SST will be providing a wealth of information that will literally turn every field properly tested, harvested and analysed into a multifaceted research plot.

2

Status of Precision Agriculture

Precision agriculture is a suite of technologies rather than a single technology. Components of that suite of technologies currently in use include GPS receivers; GIS data bases; variable-rate application equipment for seed, fertilisers, and pesticides; grid soil sampling; low-volume irrigation; yield monitors; sensors for soil fertility and weed populations; and remote sensing imagery. Different configurations of components of that suite will be suitable for different operations. It is therefore somewhat misleading to speak generally about the extent of adoption of precision agriculture. By any measure, however, adoption of precision agriculture technologies is extremely limited at present.

Variable-rate application equipment is perhaps the most widely used precision agriculture technology. About 1,600 flotation fertiliser-application systems, map-driven variable-rate technology (VRT) systems, and on-the-go sensor tractor-based application systems have been sold. Grid soil sampling also appears to be among the more popular precision agriculture technologies.

Computers are a central component of information-intensive agriculture. Fewer than 10 percent of the 1.9 million producers and ranchers in the United States own computers. Commercial yield monitors are available for corn, soybean, and wheat harvesting. Approximately 2,000 yield monitors (cumulative) were in use in the 1995 growing season. This figure increased to about 9,000 in 1996. Midwest corn and soybean growers are currently the major buyers

Gradual Diffusion of New Technologies

There is a large body of economic and sociological literature on technology adoption in general, as well as on information-intensive agricultural technologies such as computers, integrated pest management (IPM), low-volume irrigation systems (drip, center-pivot, and other sprinkler systems), and the California Irrigation Management Information System (CIMIS), which combines localised weather information with field-level soil moisture monitoring to improve irrigation management. The history of the diffusion of these earlier information-intensive technologies offers insights into the likely prospects for current precision agriculture technologies.

It has long been recognised that new technologies diffuse gradually. Technological diffusion typically follows an S-shaped path over time. In the early years after a new technology is introduced, it is generally used by only a small percentage of those who could benefit from using it. As time passes, the rate of adoption tends to increase and diffusion becomes more rapid. Finally, after the majority of those who stand to benefit from using the technology have begun using it, the rate of diffusion slows again.

It is important to distinguish between two key variables characterising the diffusion process: the extent (or ceiling) of adoption and the rate of adoption. The adoption ceiling pertains to the long term, when the diffusion process approaches completion. The rate of adoption pertains to the short term, while the diffusion process is in progress. Griliches's work on hybrid corn established that both the adoption rate and ceiling are influenced by economic factors. The adoption ceiling is influenced almost entirely by economic factors. In the short term, however, the rate of adoption is influenced by factors such as learning, risk and risk preferences, information, and human capital as well as by profitability considerations.

Principal Determinants of Long-Term Adoption

Profitability is the principal determinant in the long term of adoption of new technologies in agriculture and elsewhere—technologies that are more profitable will be used. However, adoption of any farming technology is unlikely to be universal because agriculture is characterised by a high degree of heterogeneity. Farming conditions vary markedly among regions and crops because of differences in climate, soils, topography, water availability, government programmes, and other factors. As a result, the profitability of any given agricultural technology may differ greatly across regions and crops, so that producers in one region find unprofitable what producers in another find extremely profitable.

Differences in climate are arguably the major source of heterogeneity in agriculture. They lead to differences in crop productivity and thus in the long-term profitability of adopting new agricultural technologies. Differences in climate also result in differences in pest pressure that affect ceiling rates of IPM adoption. For example, the use of scouting services for cotton crops is more prevalent in the Delta states than in Texas or California, which are drier and are thus less subject to pest pressure.

Differences in land quality (i.e., soils and topography) are a second major reason for heterogeneity in agriculture and for differences in ceiling rates of adoption of agricultural technologies. For example, low-volume irrigation technologies (drip and center-pivot) increase the efficiency of water use more on land with sandy soils and greater slopes and thus are more likely to be adopted by producers operating those types of land. Adoption of conservation tillage has similarly been more widespread among producers operating more erodible land.

Differences in cost structure are frequently another source of differences in ceiling adoption rates. For example, use of low-volume irrigation technologies is more widespread in areas where water prices are higher, so that savings in water costs are more likely to outweigh initial investment costs for drip systems. The use of computer services is more widespread among dairy operators than other kinds of producers because computerisation reduces the management time and cost involved in herd improvement.

Information-intensive technologies such as low-volume irrigation or chemigation frequently increase yield. These yield increases are worth more for higher-value crops, suggesting that ceiling adoption rates of information-intensive technologies will be greater for these crops. For example, adoption of drip irrigation systems has been greater for higher-value crops, as has the use of CIMIS for irrigation management. Adoption of drip irrigation for sugar in Hawaii has been greater on farms where yield differentials between drip and sprinkler systems are higher. Adoption of center-pivot irrigation systems in the High Plains has been greater for corn than wheat or sorghum because corn yields are more responsive to irrigation.

New agricultural technologies may spur expanded production of specific crops or livestock. Thus, the number of potential adopters of these new technologies may exceed the number of producers currently in the industry. Similarly, the extent of adoption (i.e., as measured in acreage on which the technology is used) may exceed the current size of the industry. New farming methods have frequently allowed expansion of cultivation into areas that were previously considered unsuitable. Examples include irrigated corn production on the sandy soils of the High Plains through use of center-pivot irrigation and expansion of orchards and vineyards onto hilly areas in California through use of drip and sprinkler systems.

New agricultural technologies may also turn out to have advantages that were largely unanticipated when they were introduced. As a result, ceiling adoption may differ significantly from initial expectations. For example, sprinkler irrigation is used in preference to drip systems for citrus in California because it provides frost protection. Frost protection is similarly an important motivation for the use of CIMIS, which was designed mainly to improve irrigation scheduling.

Determinants of the Speed of Diffusion of New Technologies

The speed at which the diffusion of new technologies occurs depends on a variety of factors, including:

- how rapidly information about new technologies spreads;
- how risky the new technology is perceived to be and how rapidly those perceptions change;

- — how rapidly producers can adapt to using new technologies, which depends in turn on education and other forms of human capital;
- — how rapidly learning-by-doing and learning-by-using occur; and
- — how rapidly producers of the technologies improve reliability, cost, and ease of use.

Less is known about the performance of newly introduced technologies because producers lack experience using them. As a result, the new technologies tend to be viewed as less productive, more costly, and riskier than established technologies. Initially, only a few producers will find using them worthwhile. These early adopters tend to have greater human capital or be more accepting of risk. Other producers may believe that the new technologies will not be profitable enough to justify the cost of adopting them (the cost of new equipment, plus the costs of restructuring farm operations and training) or may be too averse to risk to adopt the new technologies even if they appear to be more profitable (until the expected improvement in profit is large enough to outweigh the risk).

Risk aversion on the part of lenders may also prevent early adoption by limiting producers' ability to borrow money to invest in new technologies. Aversion to risk may also lead producers to test new technologies partially at first (i.e., trying them out on a field or trying out key components of a system). Larger operators may be more diversified against risk and thus be more willing to test out new technologies partially. Finally, some producers may find the new technologies less profitable than current technologies given the productivity of their existing capital stock.

As producers gain experience with the use of the new technology, either directly or through improved information, estimates of the return to using them tend to rise whereas perceptions of their risk tend to fall. Perceptions of risk and return for new technologies change over time in response to several factors.

The first such factor is the rate at which information about a new technology spreads to potential users. The classic model treats the process of information flow as proceeding largely through word-of-mouth, so that it mimics the spread of an epidemic. The greater the potential profitability of a new technology, the faster information about it is disseminated. In agriculture, technology-transfer programmes within cooperative extension services and marketing by equipment manufacturers, dealers, and consultants all work to spread information about new technologies.

Potential profitability affects the flow of information and thus the speed of diffusion through its effect on marketing efforts. Technologies that promise greater increases in returns tend to be marketed more aggressively, so that information about them reaches potential users more rapidly. Once the profitability advantages of new technology are well established, lenders may also promote adoption of it to their clients.

Marketing can also have negative effects on the spread of accurate information. Overzealous marketing can create inflated expectations about performance. Producers disappointed by the gap between promise and performance in initial trials may come to underestimate the potential profitability of new technologies. For example, the gap between inflated claims about drip irrigation in the 1970s and actual performance of drip systems produced an adverse reaction among producers that significantly retarded the spread of this technology.

The second factor in changing perceptions is the user's direct experience with the technology. As producers use the new methods they may come to perceive them as more profitable and less risky. In some cases increased familiarity alone changes perceptions. The spread of computer use in agriculture illustrates this learning-by-using. Producers began with simpler programmes (spreadsheets and word processing) and then branched out into more sophisticated management systems after becoming more familiar with computers in general. In other cases, user experience with new technology can lead to improvements that actually make the technology more profitable to use, rather than simply changing perceptions about profitability. For example, the performance of early drip irrigation systems was often poor because drip lines clogged. User experience with clogging problems led manufacturers to redesign the drip systems, which led in turn to accelerated diffusion of these systems.

Third, the cost of producing and installing equipment for the new technology frequently falls over time because of improvements in technology design and cost reductions resulting from production experience of the manufacturers (i.e., learning-by-doing). Lichtenberg, for example, showed that increases in irrigated soybean and corn production in the Northern High Plains were due in part to falling costs of center-pivot irrigation systems. Initially, only the most skilled producers were able to produce winter vegetables. Over time, these producers developed standard methods for vegetable cultivation under Israeli conditions, which less-skilled producers were increasingly able to use.

Finally, the profitability of a new technology relative to existing ones may rise over time as the existing capital stock ages and becomes less productive. As producers replace their existing equipment, they tend to invest in equipment embodying the new, more productive technologies.

Information-intensive technologies may have high fixed costs, either in terms of equipment purchases or in terms of acquiring the skills necessary to use them. The economic theory of investment suggests that adoption of such technologies may be discontinuous over time, because uncertainty about future conditions makes waiting a preferred option. In some cases rapid diffusion has occurred after periods of extreme conditions resulted in substantial alterations in expectations about future prices.

It is important to recognise that producers are buying the services of the technology, not the equipment that embodies the technology. Purchasing those services does not necessarily require purchasing the equipment, although equipment purchase may be the most economical means for some. There are many other ways of packaging and selling those services, such as equipment rental, custom operations for hire, or consultant services. For example, in early years, sellers of drip irrigation systems rented equipment to producers rather than requiring its purchase. As noted above, scouting services are typically delivered by crop consultants, in the form of pest management recommendations. Laser levelling, which requires investment of substantial sums in machinery, is typically provided by custom service similar to grain harvesting.

Effective and profitable use of information-intensive technologies such as precision agriculture requires high human capital, which thus plays a critical role in the technology diffusion process. The greater the availability of such human capital, the more rapid the spread of technology. Producers with greater human capital are more likely to adopt new technologies, as has been documented for computers and computer software and IPM.

Human capital can consist of producers themselves if they have the necessary education or computer training, as documented in the studies mentioned above. Alternatively, producers may rely on consultants who make recommendations derived from the use of these new technologies. Operations that are sufficiently large may hire specialised workers to provide these services. For example, in 1994, 29 percent of cotton producers performed field scouting themselves, 48 percent relied on consultants for scouting services, and 5 percent used employees. Similarly, diffusion of precision agriculture has been more rapid in areas with larger numbers of dealers and consultants to provide advice and service.

In the short term, lack of access to consultant services may limit the speed of diffusion of new technologies. Over time, however, consultant and training services should become more available in areas with high demand, as chemical and equipment dealers and independent consultants respond to opportunities to start up or expand business operations. The public sector frequently also plays an important role in training and thus augmenting the supply of human capital in agriculture. For example, land grant universities have provided the basic training in crop and pest management sciences underlying IPM, and extension agents in many states provide specific pest management training needed to implement IPM.

The diffusion of precision agriculture technologies may proceed unevenly over time. Like computers, precision agriculture initially is likely to require significant learning, both by users and by equipment suppliers. Time may be needed for these learning effects to improve performance, reduce costs, and increase reliability sufficiently to make this technologies attractive to large numbers of producers. Precision agriculture requires significant supporting infrastructure in the form of skilled labour, software development, and hardware availability—all of which may take time to develop. Moreover, some precision agriculture equipment is costly, so that potential users—whether producers or crop consultants—may prefer to wait until uncertainties about costs and reliability have been reduced sufficiently to make investment attractive.

Long-Term Potential of Precision Agriculture

The long-term potential of precision agriculture depends on its profitability (i.e., the extent to which benefits, such as of increases in yield and savings in input costs, outweigh the cost of purchasing the services). Available evidence indicates that the cost of many precision agriculture services is modest, especially when spread over sufficient acreage, as can be done by a custom applicator. These costs are comparable to the fees charged for insect and weed scouting on major crops, which are about $3 to $6 per acre. Remote sensing imagery similarly has a modest cost when spread over sufficient acreage. Raw satellite imagery data can cost as much as $80,000 per season, but data providers can sell processed images for specific fields for as little as $7 to $8 per acre (Lamb, 1996). By way of comparison, fertiliser, lime, and gypsum costs for corn in the United States averaged $46 per acre in 1994, custom operations averaged $10 per acre, and total variable cash expenses averaged $147 per acre. Capital replacement on corn averaged $33 per acre.

The central characteristic of precision agriculture is the use of detailed information to reduce the impact of heterogeneity of production conditions on output by allowing producers to calibrate inputs according to conditions at the subfield scale. For example, growers can combine tensiometer information on soil moisture at subfield levels with prediction of evapotranspiration derived from weather forecasts hourly or daily, so that producers can vary irrigation water application to match water demand at the subfield level. Similarly, variable-rate applicators combined with fertility mapping allow producers to vary fertiliser application rates in response to natural variations in soil fertility.

Calibrating inputs according to conditions at subfield levels is likely to result in increased yields. For example, diminishing marginal productivity of nutrients suggests that yields obtained under variable-rate fertiliser application will generally exceed those obtained under uniform application calibrated according to average soil fertility. Within variable-rate application, areas within a field having higher than average fertility should receive lower fertilisation than under uniform application, while areas having lower than average fertility will receive higher fertilisation than under uniform application. Yields in areas with higher-than-average fertility will be lower under variable-rate application than under uniform application, while yields in areas with lower-than-average fertility will be higher. As long as low-fertility areas account for a sufficiently large share of the field, variable-rate fertiliser application will result in an increase in yield for the field as a whole.

Precision agriculture is likely to have a greater profitability advantage than current farming methods for higher-value crops because yield increases resulting from more precise input application (should they occur) are worth more in such cases. Recent commercial developments also support the notion that interest in precision agriculture is likely to be greater on higher value crops. For example, sugar beet quality is sensitive to nutrient application and can suffer from both nutrient deficiencies and excesses. American Crystal Sugar Company, the largest sugarbeet producer in the United States, is owned by more than 2,000 North Dakota and Minnesota sugarbeet growers. This cooperative has increased its grid soil sampling in Minnesota's Red River Valley from 13,000 acres in 1995 to 130,000 acres (about 35% of the company's sugar beet acreage) in 1996.

Precision agriculture technologies may have important ancillary uses. For example, the purchase of remote sensing information on crop growth may permit closer marketing ties between producers and grain elevators by improving yield forecasts. As a result, even producers who experience relatively little within-field yield variability may find their share of the increased returns from improved marketing sufficient to make the use of such information profitable.

Precision Agriculture Services

Precision agriculture services may be provided directly (i.e., for custom hire) or may be purchased in the form of hardware and software products embodying those services, or as combination of products and services. In all likelihood, both forms of obtaining precision agriculture services will coexist, with the exact combination depending on such factors as the size of the operation, the technical expertise of the operator, and the density of the local market for services.

New services related to precision agriculture could arise independently but are more likely to evolve from services now offered by crop consultants and input suppliers. Nowkin estimates that crop consultants provide knowledge-based services on 16 percent of U.S. crop acreage, including 53 percent of cotton and vegetable acreage, 21 percent of corn acreage, and 13 percent of soybean acreage. A study of service offerings by 228 Wisconsin agricultural chemical dealers showed that although precision agriculture services accounted for only a small portion of total gross revenues, the percentage of firms offering these services was increasing.

Precision agriculture services can be provided to the producer through traditional distribution systems or by consultants. If services are provided directly to the producer, a consultant could design, integrate, and install a precision agriculture system (i.e., a combination of a global positioning system [GPS], yield monitor, and geographic information system [GIS]) for the operation, much as a computer consultant assists a small business with its computing needs. Alternatively, traditional input suppliers could be the primary customer, acting as general contractors for specific skills, expertise, and services provided by the technology consultant.

If they have access to sufficient financial capital, technology consultants could invest in the computers, software, and equipment that suffer from rapid technological obsolescence. If the consultant leases precision technologies to the input supplier, then the producer can have access to the most advanced equipment. In return for their investment, the technology consultant can depreciate the capital costs over a larger acreage base. A variation of this scenario could be a technology consultant who either competes with or supplements the services of accountants, marketing consultants, or other financial assistants in automating all of the farm's electronic accounting and record-keeping functions.

Precision agriculture services will likely be provided locally. For example, computer software such as nutrient or pest management recommendation models will need to be adapted to local (even farm-level) conditions (which tend to vary substantially). Firm size will likely be small because the number of producers in a local area would be limited. However, hardware and software support channels for the service providers could be concentrated at the wholesale level.

Provision of Precision Agriculture Products

An alternative pathway of development of precision agriculture assumes that producers and their crop consultants will buy software and hardware products to implement precision agriculture, thus limiting their purchases of services. That is, the expertise and knowledge needed for precision farming will be embodied in machinery and software to a greater extent than in the services provided by consultants. Integrating the various hardware and software components into one system would provide a seamless flow of data, culminating in either a recommendation or the presentation of alternatives to the producer.

Several manufacturers have developed integrated turnkey systems that combine GPS receivers, GIS software, crop yield monitors, and VRT hardware into precision agriculture systems. Object-oriented software modules could facilitate development of products, including software for mapping and decision support. Companies providing precision agriculture products may not need to be located in most communities. Information technology companies could be concentrated in technical centers such as Silicon Valley or near universities where technical expertise is readily available. However, hardware and software support channels would need to be widespread and located at the retail level.

Moreover, both software and hardware will likely need to be customised to match local conditions, creating a demand for locally based customising expertise. This expertise could be provided through traditional farm supply sources, such as machinery dealers (especially for proprietary lines). Independent consultants may also prove economically viable in the market for hardware and software customisation services, much as they have in various aspects of computer hardware and software sales, support, and training in other industries. Firm servicing the product market could be large in size and could dominate markets, especially if developed by or acquired by large industry leaders.

Mixture of Products and Services

In reality, precision agriculture is not likely to develop purely as a service or as a product but probably will evolve as a combination of services and products. For example, individual producers may purchase and use some precision agriculture hardware and software themselves. Other producers could hire consultants who use the same or similar hardware and software. Still others may purchase the hardware and hire consultants to analyse the data and make recommendations. For example, growers with medium-sized operations and a high level of technical expertise will likely purchase a larger share of precision agriculture services in the form of hardware that they operate themselves. Part-time operators will likely purchase more custom-hired services because of constraints on their time and capital. Very large operators may both purchase equipment and hire full-time employees with the requisite technical expertise.

Potential Effects of Precision Agriculture

Three potential effects of precision agriculture have received public attention. First, widespread adoption of precision agriculture may affect employment opportunities in rural communities. Second, precision agriculture may affect the structure of agriculture, particularly the distribution of farm sizes. Third, more efficient and precise use of fertilisers, pesticides, and other purchased inputs may alleviate environmental spillovers from agriculture.

Effects on Rural Employment

Precision agriculture represents an increase in specialisation, that is the division of labour, involved in agriculture. In adopting precision agriculture techniques, farmers substitute purchased information services (in forms as varied as consultant services, specialised equipment, and computer software) for some or all of their own personal-observation-based information collection and analysis. Diffusion of precision agriculture is thus likely to result in increased employment in agricultural support services, for example, equipment sales, computer software development, customisation of equipment and software, and consultant services.

The extent to which such increased employment occurs in rural areas depends largely on the combination of direct services and turnkey products through which precision agriculture services are delivered. The greater the extent to which growers purchase precision agriculture services directly, the greater will be the demand for locally based skilled technical labour such as crop consultants and computer software developers and customisers. Any increases in employment in the production of turnkey precision agriculture products are more likely to occur in areas where equipment manufacturers and software developers are currently located, which are largely non-rural.

Increases in rural employment caused by the spread of precision agriculture are likely to be modest. Farm-related employment is presently quite limited. The farm sector as a whole provides an estimated 1.7 million jobs, or 1.3 percent of total U.S. employment. Precision agriculture is unlikely to generate substantial additional labour demand. Some precision agriculture equipment (i.e., variable-rate applicators) replaces other forms of equipment; neither the manufacture nor the sales of such equipment will require expanded employment. Sales of turnkey products will likely not require increases in sales personnel, although more skilled personnel may be needed to service such products. Software development and customisation do not generally require extensive increases in employment.

Above all, the size of the market for precision agriculture products (both equipment and services) in the United States is limited. In 1994, there were an estimated 1.4 million households containing farm operators or managers in the United States and an estimated 2.1 million farm operations (Economic Research Service, 1996). But only a small percentage of those farm operations generate significant demand for farm equipment and services. In 1994, for example, 122,000 farm operations accounted for over 51 percent of total cash expenditures in U.S. agriculture and almost 58 percent of net cash income.

An additional 224,000 farm operations accounted for over 22 percent of cash expenditures and 26 percent of net cash income. Thus, a total of 346,000 operations accounted for 74 percent of total farm cash expenses and 84 percent of net cash income, indicating a limited customer base for precision agriculture equipment and services. The sales volume of agricultural equipment like that used in precision agriculture is similarly limited. In 1990, less than 25,000 pieces of crop harvesting equipment and about 50,000 power sprayers and dusters were shipped in the United States.

In 1991, farmers spent $604 million on planting and fertilising machinery, $2,158 million on harvesting machinery, $167 million on cultivators and weeders, and $299 million on sprayers and dusters of all kinds—a total of $3.2 billion on the kinds of machinery that precision agriculture is likely to affect. A market this limited in size is unlikely to support increases in employment that are large on a national scale, even if some rural communities eventually do experience substantial employment growth. As noted above, any gains in employment with precision agriculture will likely require workers with greater skills who would earn higher wages than is typical in rural communities.

Effects on the Structure of Farming

Some observers have argued that technological change, particularly the diffusion of information-intensive technologies such as precision agriculture, may increase concentration and further reduce the number of family farm operations. In an industry as competitive as agriculture, farm operations using less-profitable technologies are unlikely to remain in business. Producers using less-productive management approaches and supporting technologies may find it difficult to cover their production costs. Some producers who have difficulty adapting to new, technologically advanced farming methods may find it more lucrative to sell or rent their land than farm it themselves.

A hallmark of precision agriculture is the application of information to improve farm decision making. Precision agriculture thus tends to confer a competitive advantage on growers more at ease with or more highly skilled in the use of modern information technologies, who are likely to be younger and have more formal education. The spread of precision agriculture may accelerate older, less well-educated farmers into retirement from agriculture. It is important to recognise in this connection that exit from farming does not necessarily imply hardship, at least for farmers owning land. If precision agriculture is more profitable, its use should lead to increases in the value of land. Those exiting farming by selling or renting their land should thus be able to claim at least a share of the increased profit generated by precision farming methods.

It is also possible that the costs and benefits of precision agriculture vary systematically across farm size. There is concern in particular that the net benefits of precision agriculture are greater for larger operations. It is difficult to assess how the net benefits of precision agriculture vary across farm size because precision agriculture is a suite of technologies whose specific components differ across farms. In general, however, there does not appear to be an unambiguous size bias in precision agriculture or similar technologies.

Some considerations suggest that smaller operators will benefit less from precision agriculture than will larger operators. Precision agriculture substitutes somewhat for manual information collection and processing by the individual producer. Producers whose operations are small enough to allow them to know all parts intimately may benefit very little from automating information collection. Precision agriculture allows producers to reduce heterogeneity in farming operations, which permits improved profitability through reduced input expenditures, increased yields, or both. An alternative strategy to reduce heterogeneity, however, is to divide the farm into fields that are relatively homogeneous, permitting the same sorts of gains from uniform management. Producers whose operations are of a size that permits such intensity of management are unlikely to gain much from automation of information gathering and processing.

Other factors suggest the opposite—that smaller operators may gain more from precision agriculture than larger ones. The number of small farm operations is large. Automation may be more valuable to smaller, part-time producers, who operate under tight labour constraints and may otherwise be unable to increase their monitoring of field conditions. Automation may also narrow differentials in management by allowing smaller operators to access the same sophisticated strategies as larger operators. However, the current competitive advantage of some smaller operators—being able to focus on individual plants in high-value crops—may be lost when the adoption of information-intensive approaches enables larger operators to do the same.

There is similarly no clear-cut evidence that precision agriculture exhibits substantial economies of scale, so that larger operators find precision agriculture less costly on average (i.e., per acre) than smaller operators. Equipment such as variable-rate application units, GPS receivers, and computer hardware may decrease average costs because their capital costs can be spread over larger acreage. Experience with similar information-intensive technologies fails to support the notion of a consistent size bias. As noted, the per-acre cost of scouting (a central component of IPM) is comparable to the per-acre cost of many precision agriculture technologies.

There is no consistent relation between farm size and the use of scouting. For example, Napit et al. examined factors influencing the use of scouting throughout the United States. They found that the value of farm sales was positively correlated with scouting on corn in Indiana, apples in New York, cotton in Texas and Mississippi, and alfalfa in the Northwest, but not correlated with scouting on soybean fields in Virginia, peanuts in Georgia, apples in Massachusetts, or tobacco in North Carolina. Fernandez-Cornejo et al. found that larger vegetable producers in Florida and Texas were more likely to scout, but that farm size had no significant relationship to scouting on vegetables in Michigan, a finding they attributed to greater familiarity with IPM in Michigan. Fernandez-Cornejo et al. also found that farm size was not correlated with scouting for either insects or diseases for tomatoes in Florida. Additionally, Harper et al. found that neither the number of rice fields operated nor the acreage per field influenced adoption of sweep nets for monitoring stink bug pests on rice in Texas.

Low-volume irrigation involves much higher investment in equipment than most precision agriculture technology. Moreover, low-volume irrigation exhibits some economies of scale (i.e., pump size does not necessarily increase as farm size increases), suggesting that the returns to investment in this technology are likely greater for larger operations than smaller ones.

Nevertheless, the literature gives no clear indication that larger farms are more likely to adopt this technology. Adoption of low-volume irrigation technologies has been more prevalent on larger sugar plantations in Hawaii, larger tree crop operations in the southern San Joaquin Valley of California, and larger farms on the west side of the San Joaquin Valley. By contrast, neither adoption of sprinkler irrigation in the western United States nor adoption of modern irrigation technologies in citrus groves in Israel shows a significant relation to farm size.

For both low-volume irrigation technology and scouting, farm size appeared to be positively correlated with adoption in areas which were in the early stages of the technology diffusion process and not correlated in areas in the later stages of the process. For example, low-volume irrigation has been used extensively in Israel for years, whereas interest in low-volume irrigation in California became widespread only during the drought of the late 1980s. Taken together, studies of information-intensive technologies such as low-volume irrigation and scouting show no consistent relationship with farm size over the long term, although they do provide some evidence that larger operators are more likely to be early adopters.

Differential access to precision agriculture products and services marketed by consultants is also a controversial issue. Some argue that consultants offer their services preferentially to owners of larger farms. Through its workshops, the committee heard from consultants who indicated that their services were offered on a per-acre basis, suggesting equal access of services to large and small farms. Few data indicate that consultants price their services according to farm size. However, it has been argued that input suppliers in competitive markets may offer their service at no charge or at a discount, so that growers in geographic areas under-served by input suppliers have less incentive to adopt precision agriculture technologies.

It is important to put these theoretical concerns in the context of current trends in the structure of farming. The number of farms in the United States has declined consistently over time. But the rate of decline has slowed considerably in recent years. Between 1969 and 1982, for example, the number of farms in the U.S. fell from 2.7 million to 2.2 million, an overall 18 percent decline, corresponding to an annual average rate of 1.4 percent. Between 1982 and 1994, the number of farms in the U.S. fell to under 2.1 million, a total decline of 8 percent, corresponding to an average annual rate of decline of 0.7 percent.

Overall, then, the spread of information-intensive agricultural technologies (i.e., computers, drip irrigation, IPM, precision agriculture) over the past 10 to 15 years has coincided with a strengthening of the commercial farm sector, especially its moderate-sized component, at the expense of part-time operations and operations too small to be commercially viable.

Processors and Vertical Integration

Increased information about market trends may lead to greater vertical coordination (i.e., contracting of designer crops and greater vertical integration in livestock industries) and in some cases removal of intermediaries. Producers may lose independence regarding planting decisions if commodity buyers (processors) can specify product characteristics (sugar content, size, etc.) and expect their demands to be fulfilled with precision technologies. Some find such increased specialization distasteful, arguing that increased reliance on information generated by consultants, input suppliers, and hardware dealers necessarily reduces the independence of producers.

Environmental Implications

Precision agriculture may create a win-win situation by simultaneously improving farm profitability and reducing negative environmental effect of agriculture. Potential improvements in environmental quality may be a compelling reason for society to favour widespread adoption of precision agriculture, but the extent to which producers adopt it depends on economic savings from more efficient input use, not on environmental impact.

Most of the enthusiasm for precision agriculture off the farm can be traced to the good environmental sense inherent in a concept that matches input application to plant needs. Precision agriculture could be a more disaggregated version of the kinds of best management practices already recommended at the field scale. Furthermore, precisely matching fertiliser and pesticide inputs to the capabilities and needs of a crop for small areas, and applying them exactly when needed, is a logical way to limit the amounts of these materials added to the environment.

Agricultural pollution comes from inputs that do not reach their target (water or nutrients not taken up by plants, herbicides that do not come into contact with weeds, etc.). The inputs that contribute to pollution are thus wasted from a productivity point of view. Calibrating input usage more precisely should increase the percentage of applied inputs taken up by crops, thereby reducing economic waste and release into the environment. Adjusting fertiliser application rates to match variations in soil fertility at the subfield level should result in less excess nutrients and thus less runoff and percolation to groundwater. Applying pesticides to only those parts of fields with sufficient pest pressure to warrant treatment should result in less excess pesticide in the environment. Calibrating water application rates to match plant uptake rates (i.e., through low-volume irrigation) should reduce excess water applications and thus mitigate drainage problems.

The extent to which more precise input applications might reduce environmental conditions is by no means clear. A number of field-level experimental studies indicate that variable-rate nitrogen fertiliser application can reduce the nitrogen application rate needed to attain given yield levels. But these decreases in application rates have sometimes resulted in little or no discernible decreases in post-harvest soil nitrate concentrations, suggesting few if any, consequent benefits in terms of ground or surface water quality improvement.

Synergy between VRT and biotechnology could also lead to environmental improvements. Biotechnology firms and seed companies are racing to manipulate the genetic makeup of commercial crops to include tolerance to herbicides. Beginning in the late 1980s, genetically engineered innovations such as imidazolidinone-tolerant corn, sulfonylurea-tolerant soybean, and Buctriltolerant cotton varieties were developed. The natural insecticidal properties of *Bacillus thuringiensis* are being incorporated into corn, soybean, and cotton varieties. Although these genetically enhanced seeds can confer economic and environmental benefits when used on a whole field, they are likely to be more effective when applied on a precision basis.

For example, if there are yield penalties associated with some of these varieties, they can be variably planted in areas with high weed infestation or areas in which sensors indicate higher organic matter that could be associated with a greater need for pre-emergence herbicide application. Precision application of seed enhanced by *B. thuringiensis* could reduce problems with development of resistance that would be accelerated if whole-field application were used. Variable-rate seeding may help match plant populations to soil conditions, and precision seeding with enhanced varieties can be used to apply seed enhanced with *B. thuringiensis* or herbicide tolerance in potential pest problem areas.

Apart from empirical studies, it is possible to envision situations in which precision agriculture both mitigates and exacerbates potential environmental problems associated with crop production. For example, variable-rate fertilisation may counteract potential yield differences between steeper shoulder slopes and shallower back and foot slopes that were unfertilised or conventionally fertilised. Increased soil cover, obtained on steeper slopes through VRT application, could reduce soil erosion, but increased application of nitrogen could increase losses to the environment if yield-limiting factors reduce nitrogen uptake. In another example, areas with droughty soils caused by rapid percolation may have lower levels of soil nitrogen because of greater leaching losses.

VRT nitrogen application could exacerbate leaching if additional nitrogen is applied to counteract losses in these soils, or could mitigate losses if nitrogen were applied in greater synchrony with nitrogen uptake. Uncritical use of precision data indicating higher yield potential in certain parts of a field could lead to higher input use, especially if recommendations are formulated without any interest in reducing chemical use. Precision agriculture may make it economically attractive to expand production into marginal lands, which may create new environmental problems. For example, the spread of drip irrigation exacerbated groundwater depletion problems because it facilitated the expansion of tree crop production onto hillsides rather than substituting for less efficient gravity irrigation methods in flat areas.

Most studies of the environmental impacts of precision agriculture have concentrated on field-level effects. But the environmental impacts that matter most are generally those that occur in ambient pollutant concentrations. Unfortunately, there is little empirical evidence currently available that precision agriculture actually reduces delivery of pollutants to surface and groundwater and the atmosphere relative to conventional techniques. Moreover, there is good reason to believe that field-level effects do not scale up readily to ambient impacts. Just as with field-level management for potential environmental problems, the effect of reductions in erosion, residual nutrients, and pesticide applications achieved through precision agriculture focused on ambient surface and groundwater quality at the subfield level depends on the position of the field within the watershed and relative to important aquifers.

Generally, reductions occurring at great overland distances from streams and lakes, or away from aquifers contributing to wellheads, will have smaller contributions to environmental improvement than similar reductions occurring closer to these water resources. Spatial patterns of farming activity within the watershed may have more of an impact on environmental quality than do improvements in environmental management within farm fields.

It is also worth underscoring that environmental improvements by themselves do not generally constitute an incentive for growers to adopt precision agriculture. In the long term, potential environmental improvements will constitute an economic incentive for adopting precision agriculture only in areas where producers bear at least a share of the costs of agricultural pollution. The problem of drainage in the San Joaquin Valley, California, is a case in point. As long as producers were able to dispose of drainage water at low cost into the San Joaquin River and Kesterson Reservoir, they had little incentive to lessen drainage problems by improving irrigation efficiency. After regulatory action resulted in severe restrictions on the use of these low-cost disposal outlets, producers began using low-volume irrigation equipment, levelling fields, shortening irrigation runs, and making other improvements in irrigation efficiency to reduce drainage accumulation (Dinar et al., 1992).

The use of precision agriculture data for environmental monitoring is also potentially a source of contention. The vast amount of site-specific data generated by precision agriculture methods may help allay regulators' concerns about the sources of environmental problems associated with agricultural production in particular areas. However, inappropriate use of such detailed data by regulators can be highly damaging to producers' interests, and fears of such misuse will undoubtedly make producers reluctant to share such data with government agencies. Ways must be found to cooperatively use data generated by precision agriculture for environmental improvement while safeguarding against abuses or inappropriate uses.

Precision data on nutrient and pesticide soil concentrations and application rates could help producers demonstrate that chemicals were applied in a legally and environmentally sound way, reducing potential liability for environmental problems. Such data could just as easily be used by regulators to prove the opposite. Decisions about data ownership and access will determine where the burden of proof lies in use of such data for environmental purposes.

3

Precision Farming Technologies

Precision farming (PF) technologies hold promise of advantages both for those farmers who adopt this technology and for society. There is potential for increased farm profits if inputs can be allocated with greater economic efficiency across all farmed acreage. Furthermore, precision farming methods may allow risk reduction for the individual farmer if decisionmaking can be improved to diminish overall variability in farm earnings over a variety of weather and other environmental pressures. To the extent that PF allows improved economic efficiency of input allocation, society will gain from lower cost production and reduced costs of food and fiber. There also may be environmental benefits resulting from PF methods.

Precision farming is not a single technology, but rather is a suite of technologies that can be combined as a system. The farmer will choose those technology components that provide for the needs of the specific farm. Therefore, the precision farming systems employed on a sample of farms may differ significantly. Likewise, the uses made of the PF system will vary with manager knowledge, preferences, and needs.

Some PF components will need to be adopted in sets in order to make them most useful. For instance, a yield monitor often is adopted without adoption of a global positioning system (GPS). When the yield monitor is adopted alone, yield measurements will be made and displayed every few seconds as the combine crosses the field. However, without GPS these numbers cannot be geo-referenced, cannot be stored in a geographical information system, and yield maps cannot be created. Addition of the GPS system allows for these outputs as well as the possibility of other spatial analyses. The end result is much greater potential to derive meaningful decision support information from the combined GPS and yield monitor.

Variable rate application of variable inputs such as fertilisers, lime, herbicides, and seed creates the potential to manage each parcel (management zone) of land uniquely, based on its specific productivity characteristics. This allows for the potential to apply inputs in an economically optimal fashion in each zone. However, to do so requires data sufficient to predict crop response to the variably applied inputs in each zone. For variable rate application of fertilisers, such data might be derived from geo-referenced grid soil sampling, or based on other data specific to each management zone (e.g., historical yield maps, crop tissue testing, remote sensing imagery, or geo-referenced crop scouting). Variable rate seeding decisions may be based on different data, for instance soil type, slope, elevation, presence of tile drainage, planting date, and expected weather.

These spatial input prescriptions could be created by the farmer based on analyses of the farm's geographical information system database, or could be "delegated" to a consultant, the local cooperative, or to other precision farming service provider. The key point is that each precision farmer may make different precision farming decisions, based on differing types of data, and perhaps derived from different PF component technologies.

The annual cost of utilising precision farming tools depends primarily on the useful life of the equipment, software, and databases you use. The fact is, new equipment and software are rapidly being put out on the market, and if your tools become obsolete in three to four years (like other computer-based technologies), then the annual cost of use can be relatively high—from $500 to $1,000 per year.

While studies of precision farming technology have often focused on changes in crop input costs, there is also the cost of developing "human capital." These costs include workshops, training fees, time away from other work, and "wrong decisions" made while learning.

Yield Monitors

Many producers associate precision farming with combine yield monitors, the most popular component of the technology. At present, there are an estimated 17,000 yield monitors in use in North America. Of these, approximately 40 percent are equipped with global positioning system technology (GPS). Yield monitors have the capability of indicating yield (bu/ac), wet and dry bushels, total pounds, acres/hour, acres worked, and grain moisture content. This is all done while the combine is in use, and can be recorded on a memory card for later analysis, mapping, etc. All information becomes site-specific when used with a global positioning system.

GPS/DGPS Systems

GPS (global positioning systems) and DGPS (differential

global position systems) are used to pinpoint the location (geo-reference) of yield data collected, and can provide accuracy of one to three meters. However, there is an annual subscription cost to use the required, differential signal (either from a satellite or a land base unit) of $200 to $600 per year.

Computer Hardware and Software

Computer support is required to analyse the data collected by yield monitors and GPS, and supply it to you in usable formats—such as maps, graphs, charts or reports. If you are buying a new computer, select one that is as up to date as possible—as well as a good quality colour printer. Yield-mapping software can be purchased as part of an overall combine package, or it can be purchased separately from $1,000 to $1,500.

Field Scouting

Field scouting is a necessity in precision farming, just as it is in any good farm management system. When you are scouting, a portable GIS unit will allow you to identify and record the location of problems or events that will affect production—including soil differences, insect infestations, fertility deficiencies, and weed problems. Remote sensing, and satellite and infrared images can also be employed while you are scouting your fields—but they can be time consuming and costly.

Soil testing in precision farming still requires you to walk your fields to take samples. Testing is based on grid sampling or topography differences, as well as on yield differences indicated by your yield mapping. In order to ensure reliable soil test recommendations, you must be sure to take adequate samples.

GIS Software

For an even more detailed analysis of your fields, additional geographical information system (GIS) software is available that will provide sophisticated maps. However, you will need a major time commitment to efficiently utilise the capabilities of this software. Many farmers find that consultants are an option in providing GIS expertise—including equipment and interpretation.

Variable Rate Technology (VRT)

Equipment is now available—and more is being developed—that will allow variable rates of inputs to be applied to a single field. However, before this technology can really get off the ground, a method must be developed to produce reliable and extremely accurate prescription maps—which will dictate the varying rates of various inputs.

Crop prices very much affect the use of variable rate technology (VRT). The expense of VRT may be more appropriate for higher-value crops, such as beans and potatoes, as opposed to bulk cereals and oilseed crops. Higher-value crops have higher input costs, so input savings can be great enough to justify the expense of VRT. As well, the value of yield increases from high-value crops can also be a reason for investing in VRT.

Advantages of Precision Farming Technologies

Precision farming makes information and diagnosis part of your whole farm system. In the short term, yield monitoring and other site-specific information give you an improved ability to diagnose crop production problems. Yield restrictions caused by drainage, soil type, crop inputs, diseases, weeds, etc. can be quickly identified and corrected—providing you with immediate returns.

In the long term, data can be collected and analyzed over time, providing you with the opportunity to evaluate the effects of various management practices on yield. You can also compare ongoing yield variations for site-specific locations or entire fields, to determine problems such as inherent soil differences or micro-climate differences.

Precision farming technology is also useful in evaluating crop inputs, new products, new methods, etc. It can generate production comparisons for a particular field or farm, so that you can make informed management decisions on use of inputs, products, and/or methods. Evaluating these comparisons without precision farming technology is time consuming and often inaccurate.

Tools like yield maps will track trends and identify problems that can help you make changes in crop rotation, variety choices, and other important decisions. For example, resistant wild oat patches can be singled out (GIS referenced) and then isolated—which will restrict their spread, giving you time to look at new strategies, rotations, and herbicides.

Efficient Use of Equipment

Information on soil characteristics and weather can be used to plan and improve scheduling of operations, which can increase machinery utilisation rates and lower per-acre costs. As well, GPS-based guidance systems can allow operators to achieve greater field efficiency under difficult conditions. They can reduce overlap and missed applications of inputs (e.g. spraying), helping fatigued operators maintain higher field efficiency.

Risk Reduction

At the field level, precision farming provides site-specific management that can point out problems with growing conditions, thereby reducing variability in net returns. At the farm. level, precision farming information can be used to improve variety choice, crop rotation, and other agronomic practices that reduce risk. As well, information no crop growth during the season can help you make more informed market decisions.

Management of Crop Differentiation

In the future, precision technology may help farmers differentiate their production within a particular field. For example, you might segregate higher protein wheat for marketing in more rewarding channels.

In addition, precision farming technology will allow the additional control that is required when you are managing the production of differentiated products, as opposed to the production of regular bulk crops. It will allow documentation of crops conditions and control of inputs to meet the very specific requirements of these crops.

Environmental Stewardship

Precise application of chemical and fertiliser will better match crop requirements, and will prevent over-application—which can be non-beneficial to the environment. The management practices generated by precision farming technologies will promote good land stewardship.

Farming is Profitable

In order for precision farming to be profitable, the technology needs to be used in ways that fit local farming conditions. For production of bulk commodities on the Canadian prairies, profitability will depend on making the best of low-cost information. There is no doubt, however, that site-specific data bases will help improve management skills and profitability. The most profitable uses of precision farming technology are likely to be found in information system applications, diagnosis of crop problems, equipment use efficiency, risk management, crop differentiation, and process control.

4

Understanding GIS

In the real world, the landscape is composed of soil (never enough), rocks (too many), water (often too little), plants (more often weeds) and a vicissitude of biological life. In the paper world, words, numbers and graphics are used to represent these things. A traditional map is a graphical representation that accounts for geographic space (where) as well as the characteristic and condition (what).

A GIS takes the traditional map to new heights and may be described by its process, data, and analytical capabilities. The GIS process involves encoding, storage, processing and display of computerized (digital) maps.

Processing functions include:

— computer mapping;

— spatial database management; and

— spatial statistics, analysis and modeling.

These functions are descriptive and interpretive as well as prescriptive in nature. Computer mapping is a descriptive process that rapidly creates and updates map products. It provides tools for automating the cartographic process, but does not alter the inherent nature of a traditional map.

Spatial database management is used to combine and interpret sets of mapped data. It provides new procedures for *geo-query* of the data. For example, a database map can be searched for map compartments with certain requirements (i.e., low pH values in a certain soil type), then produce a map locating these areas. The spatial modeling capabilities extend the descriptive and interpretive uses of maps by deriving new information through the relationships among existing maps. For example, template maps can be "summarized" for typical characteristics (such as average slope for each mapping compartment), and the result can be added as a new column to the database.

Part of the revolution in GIS simply involves *digitizing* (converting ink to numbers) familiar maps. However, GIS goes beyond just a new form of maps. It is a powerful set of new tools for *map-ematical* processing of these data and modeling complex systems, such as crop production—describing, interpreting, predicting and prescribing the potential.

Data Formats

Traditional maps are graphical abstractions containing inked lines, shadings and symbols to locate landscape features (fig. 1). *Points* represent dimensionless features, such as wells. *Lines* depict features with length, such as streams or waterways. *Areas* (or *polygons*) have length and width, such as agricultural fields. Add depth and areas are extended to *volumes*, such as the entire body of water in a pond. Hypervolumes continue to extend the dimensions beyond simple geometry, such as considering time (mapping the rising and falling levels in a body of water throughout a year).

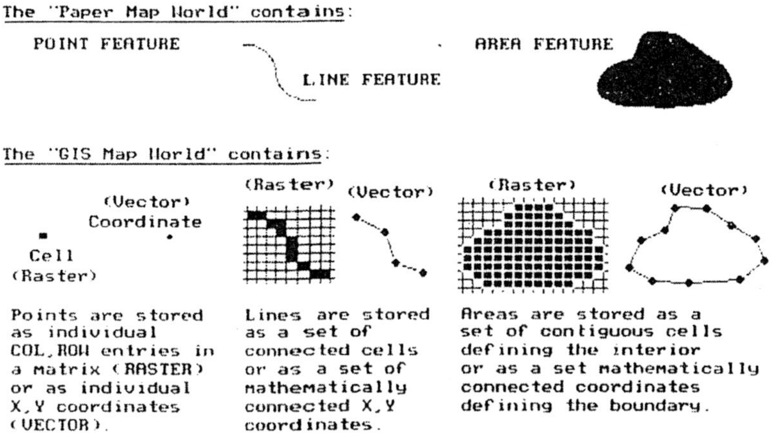

Figure 1. Basic map features used in GIS.

Most maps are composed of points, lines and areas. For example, a typical field map may identify an irrigation gate as a dot, an irrigation ditch as a squiggle, and a ponding area as a blob. A GIS can produce a similar graphic representation, but it doesn't store the data that way.

In a vector-based GIS world, map features are represented by x, y coordinates. Points are identified as a single coordinate pair. Lines are a connected set of points (like a connect-the-dot picture). Areas, such as soil type units, are identified by a list of coordinates defining the line segments forming their borders. The fundamental unit of a vector system is the point. A line is constructed by connecting individual points. An area is constructed by a line that connects to itself forming a border around the area of interest. The data structure and mathematics defining the implied line segments is extremely complex.

The basic structure of points, lines and areas, however, is very familiar as it mimics the inked markings on traditional maps.

A less familiar format, raster, uses an imaginary grid of cells (similar to a spreadsheet layout) to represent the landscape. Points are individual column/row entities (grid cells), lines are a set of connected cells and areas are stored as all the cells within the interior of each map feature. Although this data structure has several advantages, it has a major disadvantage—lack of spatial precision. If a stream passes through an acre grid cell, the whole cell is identified as containing the stream. You don't know if it's at the top or bottom or wiggles several times through the center of the cell.

At this juncture, it is important to note that vector systems "precisely locate" map features, whereas raster systems "statistically express the occurrence" of features. Generally speaking, vector is better for computer mapping and spatial database management processing functions, while raster is better for spatial modeling of the relationships within and among mapped data.

In part, this distinction arises from the recognition of a fourth basic map feature—*surfaces*. While points, lines and areas form discrete spatial entities, a map surface depicts a continuously changing gradient. Elevation, for example, is constantly changing throughout geographic space. While representing elevation as a series of discrete contour lines may be familiar, it is an inaccurate representation (a hold-over from manual drafting). Similar arguments hold for the complex spatial distributions of crop yield and soil nutrient levels, as well as a multitude of other farm conditions affecting crop production.

Though less precise at depicting discrete features, raster systems are ideal for representing map surfaces. They partition geographic space into consistent spatial units (grid cells) that are analogous to sample plots—it's just that there is a sample plot everywhere. In techy-speak, raster's contiguous set of grid cells "sample" the continuous spatial distribution of each variable and maps its spatial autocorrelation.

Also, analysis of the spatial dependency among maps is simplified by using a consistent "sampling" grid. Since the grid cells align in each map, the computer simply retrieves the necessary data stored for each cell location. A vector-based system, on the other hand, must calculate the coincidence for each irregularly bounded feature with every other feature on every map—a considerable task as most map's contain hundreds of irregular features defined by a multitude of mathematically implied line segments.

Descriptive Data Links

In the paper world, the human is the link, running back and forth between a map on a wall and data about map features stored in file cabinets. If a farm cooperative needs to know which fields have certain corn varieties growing on Cohasset soil, they flip through their files and note the field numbers of the ones they're after. Then they go to the map to locate them. If they wonder what the variety and soil type is for a neighboring field or forest stand, they run back to the files.

That's a lot of work for humans, but a snap for a GIS system. In effect, the map and file cabinets are electronically linked as shown in figure 2. A common identification number (ID#) is part of the map features (where) and the thematic attribute (what) tables.

Actually, these tables are just plain old databases—one stores the *x, y* coordinates, while the other stores the information about each feature. Each row of the attribute table (termed *records*) is divided into several columns (termed *items*). This is similar to the old days when data (items) were kept on a 4-x-5 card (record) for each field.

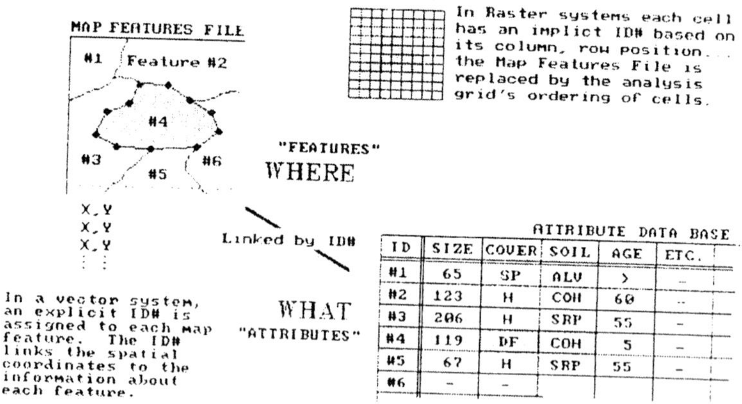

Figure 2. GIS maps linked to databases by map feature ID.

In techy-speak, a "database entry, geo-query" searches the attribute table for the items COVER = DF (for "darn fine" corn) AND SOIL = COH (for Cohasset). As shown in figure 2, map feature #4 is the only one that meets the joint condition. Its coordinates are plotted to the screen and filled with a vibrant color of your choice. The GIS searches the remaining 10,000 features (individual fields) within the county, and all of the "hits" are plotted in matter of seconds.

Conversely, a map entry, geo-query involves mouse-clicking anywhere on the map and all of the data for that field instantly pops-up. The raster world has an analogous link. Each cell has an implicit identification number (ID#) based on its column/row position. By convention, the analysis grid is ordered as you read a book, from left to right and from top to bottom. By implication, the first cell (ID# 1) is in the upper-left corner. The next cell (ID# 2) is the adjacent cell to the right. The sequential numbering continues through the last column of the first row. It then picks up the sequence for each successive row. It finally finishes with the lower-right cell. Most raster systems store their "what" information in separate files.

A few raster systems, however, store the information as one large attribute table with each record (row) indicating a grid cell and each item (column) describing a separate thematic attribute (i.e., variety, soil type, and elevation). If you think about it, the similarities between vector and raster formats should be apparent. The attribute databases, containing the "what" information, are nearly identical with the exception of the ID#'s—explicit for vector and implicit for raster. The map features file, containing the "where" information, for vector stores irregular features; for raster it is an implied analysis grid of regular cells.

Dumb Maps

When you view a map, all sorts of things are apparent. If two blue lines come together you instantly recognize it as a fork in a stream. You easily comprehend which stream networks are connected, and which are not. Your interpretation of contour lines even tells you which way water flows.

A *GIS* isn't as lucky. The relationships among the features forming a map (*map topology*) has to be contained in the data's organization. The GIS may know that a coordinate pair (x, y location) is identified with a stream, but without topology it has no idea how that location relates to all of the other map locations.

You will recall that the fundamental element of map structure in a vector system is the point, represented by an x, y pair of values. These coordinates usually relate to a standard earth coordinate system such as latitude and longitude. Although the geo-referencing is the same, there are several types of *points* (fig. 3). *Tics* are geographic control points used in registering a map. *Discrete points* are used to represent information such as wells. These independent points also are used to associate data with areas and to position map labels. *Vertices* and *nodes* are points used to construct lines and polygons. This process is similar to "connect-the-dots" drawings. Vertices are merely passed through; nodes identify the special points where more than two lines meet.

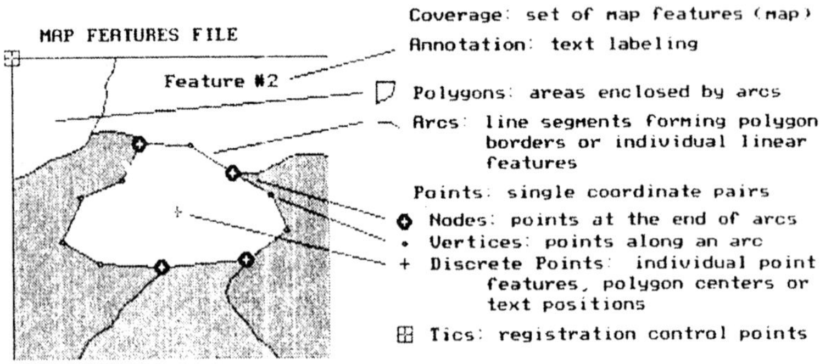

Figure 3. Basic elements of vector data structures.

The set of all line segments (termed connected vertices) between two nodes forms an arc. For linear features, such as a stream network, the GIS data structure keeps track of the connected arcs. When a stream flows into a lake, the node is tagged as an inlet. The stream node at the other end of the lake is identified as an outlet. You "see" this stuff; the computer has to be told.

Polygons are areas enclosed by arcs. Just as several points form an arc, a closed series of arcs form polygons. A *discrete point* inside each polygon (usually the *centroid*) serves as a quick link to information about it (i.e., soil, size and crop). When polygons are adjoining (i.e., soil types within a field), their shared arcs are tagged with a special code. In this manner, the GIS knows the adjoining polygons, their adjoining polygons, etc.

With minimal guidance from map annotations and labels, you see all these relationships in a paper map. But the GIS just "sees" a huge pile of numbers. It must incorporate all of the relationships you see into its data structure. When it jumps into the middle of a map (technically termed a *coverage*), it has to be able to sequentially construct all of the relationships among map features. Yep, digital maps are dumb without a data structure to keep track of all the details.

Raster Features

The vector *data structure* is constructed by connecting points, lines and arcs and explicitly storing their relationships (i.e., which polygons are touching). Raster systems store the "what" information with an implicit (rather than explicit) ID# that's based on the position of the information in a data file. The entire landscape is covered by an imaginary grid of cells, the basic unit of this data structure. Like a vector's point, there are different types of cells (fig. 4). A whole cell contains a single map characteristic throughout its interior (i.e., a particular soil type). A partial cell contains a mixture of characteristics (i.e., portions of two soil types) or an individual characteristic (i.e., a road or spring) that is too small to cover the entire cell.

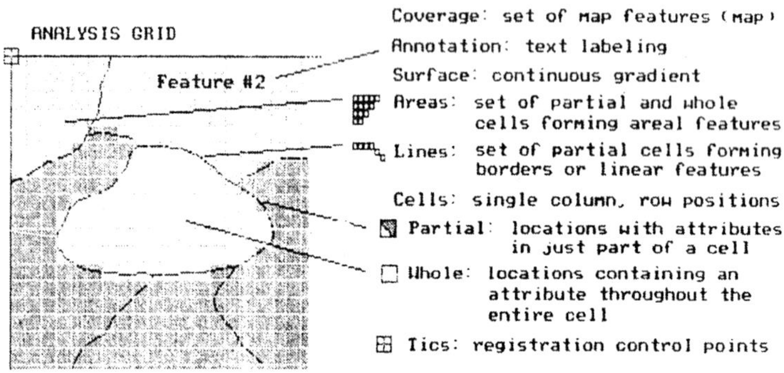

Figure 4. Basic elements of raster data structures.

It's the partial cells that account for the lack of precision in raster data. In this data structure, all of the spatial detail smaller than a cell gets lost. If a finer analysis grid is used, precision increases. In theory, the grid could be as fine as the x, y coordinates in a vector system, yielding identical spatial precision. However, the storage and processing demands at such a high resolution exceed the capacity of most modern computers.

Until a super computer sits on every desk, an oversized partial cell must be used to identify a single point in geographic space. Connected series of partial cells identify rasterized line features. A set of whole (*interior*) and partial (*border*) cells identify rasterized areas. To simplify things, the characteristic dominating a border cell can be assigned, thus making it a whole cell (and a whole lot easier to conceptualize and store).

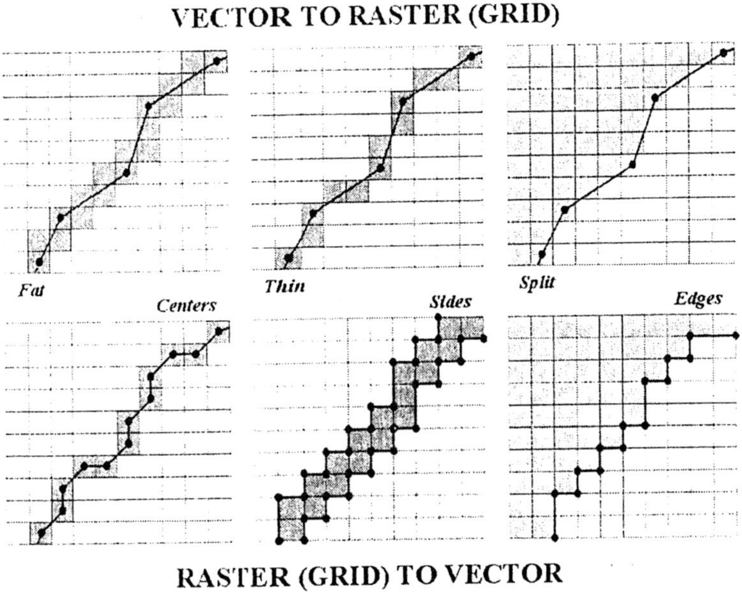

Figure 5. Approaches to conversion between vector and raster.

The raster structure allows us to extend the basic map features from just discrete points, lines and areas to map surfaces. A *surface* describes the continuous distribution of gradient data. Elevation is a classic example, at least in hilly terrain. Each *grid cell* is assigned an elevation value that typifies the elevation within its boundary. The raster format of elevation data is termed a *digital elevation model* (DEM) and contains radically different information than the traditional contour lines on a topographic map.

Atmospheric pressure, temperature and cost surfaces are other examples of this new type of map feature. In fact, *site-specific management* implies a continuous treatment of geographic space. While the infrastructure of a farm (buildings, roads fences, ditches) are best represented as vector data forming traditional maps, most agronomic data (crop yield, soil nutrients, soil moisture, nematode concentration) are inherently continuous spatial gradients best represented as raster data forming map surfaces.

Converting between Lines and Squares

A common thread in most *precision farming* analyses is the data structure employed—an imaginary grid of cells superimposed over an area of interest. The distinction between this gridded structure (raster), and the more familiar map form composed of points, lines and polygons (vector) has been discussed earlier in this topic, "Data Formats," "Data Links" and "Dumb Maps."

The most important distinction affecting data analysis is that raster data characterizes a consistent "parceling" of geographic space while the vector structure contains sets of irregular parcels. The imaginary grid of uniform cells represents an area as a continuous surface that is identical throughout a single map and between map layers. The vector structure, on the other hand, represents discrete objects that provide little information about their spatial context and none about the consistency needed for efficient spatial analysis within or among maps.

Conversion between the two data structures is easily done, but at a cost in spatial precision and informational accuracy. Transforming a vector point simply identifies the raster grid cell containing the point. Reversing the process simply uses the centroid of a cell to represent a point feature. If the cells used the translation are small, the imprecision introduced isn't much. The positioning error is a function of cell size (worst case is one-half the diagonal distance of the cell) and dramatically increases with larger cells.

Conversion of line features is a bit more complex. As shown in figure 5, two basic options hold—fat and thin. Fat raster representation uses simple geometry to identify every cell that contains a line, even if it just touches a corner. Thin raster uses a *spatial filter* to reduce the set of "fat" cells to those that best represent a linear feature.

In practice, fat cells are used for blurry features with uncertain positioning and thin cells are used for well-defined features. Some linear features (i.e., an irrigation canal) can go either way—thin to best represent the canal's positioning or fat to statistically identify all cells that have direct access to water.

Polygonal features divide continuous space into cells best representing the condition within each cell. An interior cell is easily assigned when the same condition covers the entire cell it's the edge cells that present the problems. Most procedures split the set of edge cells based on their dominant condition. Geometry is used to determine which side of a line contains the most area of a cell. Another approach uses fat cells to identify a separate "transition" category and identifies the mixed conditions within each edge cell (e.g., some soil A and some soil B).

Conversion from raster to vector is just as inexact and plagued with simplifying assumptions. With vector-to-raster conversion, precise locations are "blurred" and represented as occurring somewhere within a cell. Raster-to-vector conversion, on the other hand, takes a generalized location (somewhere within a cell) and infers precise positioning—an art form at best. The simplest approach uses the centers of the cells to extract a list of x,y coordinates that are used to position point features or a string of connected centroids to locate a line. Many systems use spline functions to mathematically smooth a line like a draftsman's french curve template.

An alternative for representing linear features "outlines" the sides of cells (usually fat cells) for a vector representation of "corridor" containing the line—in a sense, a more honest reporting than inferring precise placement from a smoothed set of connected centroids. In a similar vein, the edges of the border cells of a polygon can be outlined (and smoothed) to form a precise rendering of the implied boundary, or the sides of border cells can be identified to represent a transitional zone.

In precision farming, vector-to-raster or raster-to-vector conversions play an important role. Many of the input maps and popular map display systems (e.g., ArcView and MapInfo) utilize traditional vector data structure. However, most data analysis utilizes raster data. Conversion between the two is not an exact science and users should be aware of transformation approaches and sensitive to the interpretation of their results.

GIS Workspaces

A GIS workspace is like your desktop when you are working on a project. It has all of the information (books, maps, papers, calculator, etc.) that you need to tackle a problem. Within a computer, the data and programs you need are assembled into a workspace. The proper term for a GIS map is a *coverage* (either vector or raster format) as identified in figure 6. Three things make a GIS coverage different from a traditional paper map.

— It's digital (numbers instead of ink).

— It represents only one theme (the "what" of a map).

— It's seamless.

While a paper map composites several themes (i.e., political boundaries roads, water and soils), a GIS stores each theme as a separate coverage. In addition, each of the categories defining a coverage must be mutually exclusive in geographic space—we can't have two soil types assigned to the same location.

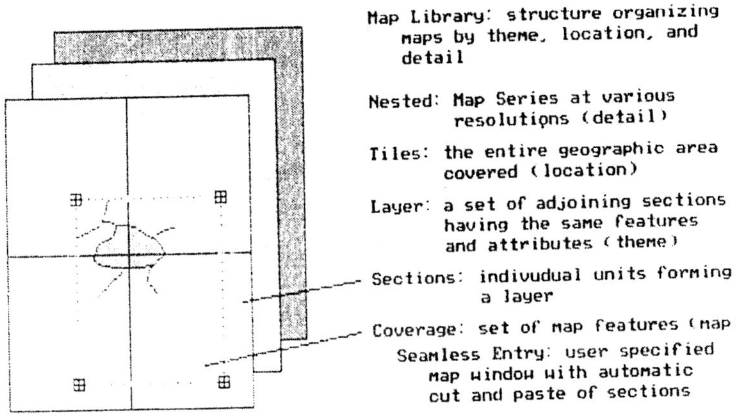

Figure 6. Basic elements of GIS database organization.

Seamless means a user can specify a set of corner coordinates and the GIS will automatically "cut and paste" data from the appropriate storage locations. This process is similar to your locating four adjoining topographic maps, identifying your project boundary on each, whacking away with your scissors, then taping the pieces together. A section, in GIS-speak, is simply the computer's means of dividing a large area into regular blocks for efficient referencing and disk storage.

A layer is a set of adjoining sections having the same features and attributes. For example, a GIS database might contain separate layers for field boundaries, drainage tiles, soil type, nutrient levels, crop yield and terrain slope. Each layer is stored separately but can be printed as a "sandwich" of inked layers. You must take great care in encoding each section to make sure that they edge match; the boundaries of features must continue from section to section. This requires precise digitizing (e.g., tracing map features into the computer) and consistent classification schemes for the mapped data.

While a *layer* describes the informational content of each GIS map, a *tile* describes the geographic area represented in each of the layers. It identifies the geographic extent covered in by the *workspace*. A extension agent's "tile" is likely to include an entire county or two, while a farmer's "tile" might cover just a few fields. In large part, the size of the tile determines the size of the database and, ultimately, the size of the computer system and complexity of the software needed to handle the data. Most farm-level workspaces easily fit in a standard personal computer environment. Larger "chunks" of land for co-op or government-level perspectives require larger computers and extensive data networks—like shoes, a "one-size" system does not fit all.

A *nested map* series contains maps at various *resolutions* (levels of detail) over the same geographic area (*tile*). This concept is particularly applicable to raster databases containing different *remote sensing* data. Frequently, a user will store two or more analysis grid resolutions of the same area—a coarse one for strategic planing and a fine one for tactical studies.

The final workspace concept ties it all together—the *map library*. A map library lists all the GIS maps in a system. The listing is simultaneously organized by location, theme and detail. In a full-feature GIS, you can specify a project area, select the maps you need, then store them in your own workspace. With a healthy understanding of the errors involved, you can transform maps at various geographic scales and projections, resample maps of various levels of detail, as well as exchange vector and raster maps.

All this has been an overload of important, but mundane and arcane terminology. Some of it makes common sense; some of it may make no sense at all. In many ways a digital map is as different as it is similar to traditional paper maps.

Table 1. Basic GIS terminology.

Spatial Data Characterization	Vector Data Model	Raster Data Model	Spatial Data Organization
Points, lines, areas	Map topology		
Cells (col/row)	Workspace		
Volumes	Points (x,y)	Whole cells	Coverage
Map surfaces	Control tics	Partial cells	Seamless entry
Hyperspace (time)	Discrete points	Rasterized lines	Sections
Vector data	Vertices	Rasterized areas	Layers
x,y coordinates	Nodes	Map surfaces	Edge-match
Map features	Line segments	Implicit topology	Tiles
Thematic attribute	Arcs		Nested maps
Identification #	Polygons		Map library
Table records	Map annotation		
Table items	Explicit topology		
Raster data			
Col/row entries			

Keep in mind that you're the intellectual superior of the GIS. You simply "see" things on paper while the computer has to organize everything in excruciating detail on disk. Although fundamentally different, you and your GIS need to be agreeable partners. Table 1 summarizes the "terms" of this partnership.

Spatial Topology

Your computer thinks of a map as simply an organized set of numbers. *Spatial topology* describes the linkage of numbers and map features. Think of it as the information that gets added to map coordinates so the computer can "see" their relative positioning. Different software allows the computer to see maps differently. A graphics package, or paint-brush type system, does a great job of coloring maps because the computer sees the map as "on-and-off dots," all assigned an appropriate color. Your eye assembles the dots into patterns, but the computer only sees a pile of independent dots. A graphics package can draw maps, but it doesn't supply the spatial topology needed for analysis.

A *mapping package* offers more than "connect-the-dots" topology. It divides up the set of coordinates into separate groups of map features. It uses the same approach as *computer-aided design* (similar to .DXF files), with points (a pair of coordinates), lines (a string of connected coordinates) and polygons (a line that closes on itself). A mapping package can draw maps and keep track of separate map features.

A *spatial database management system* (SDBMS) adds another ability: using a "connect-the-dots-to-records" structure for *thematic mapping* and *geo-query*. With this approach, a CAD-type structure of map features (*spatial table*) is linked to a database containing information about each feature (*thematic table*). It uses a common ID number for each feature, which is found in both the spatial and the thematic tables.

A *thematic map* is generated by assigning colors to a range of conditions identified with one of the map themes. For example, a farm co-op can generate a corn yield map for a county by assigning colors to low, medium and high yields for each field's aggregate yield. A farmer can generate a yield map of each field by assigning colors for low-, medium- and high-yield measurements recorded by a yield monitor with *GPS* positioning.

A *geo-query* selects sub-sets of the data. With the simplest geo-query, mouse-click on any map feature and the thematic data linked to it pops up. Conversely, the thematic table can be queried, which results in a map. For example, a farm co-op can select certain farmers' fields for thematic mapping of yield. Or a farmer might ask for a map of field locations that have both high yields and high moisture content. The power of an *SDBMS* is in the ability to visualize existing data.

A GIS fills in several gaps in the computer's understanding of the relationships in mapped data. A GIS provides a topology based on "connect-the-dots to records and to spatial relationships." This relational capability separates pretty maps from GIS maps—and provides software with the ability to perform map analysis. For a computer to find its way around a map, the connections you see must be stored in the spatial table, as well as the list of coordinates (vector *x,y* or raster column/row) positioning each feature. For example, a farm co-op might be interested in deriving a map of all high yielding fields within 20 miles of a grain elevator. Or a farmer might calculate a percent-change map for a field's yield over two years, then generate a map of locations that increased (or decreased) by a certain percentage.

Generally speaking, a topologically structured GIS allows the derivation of new spatial information based on existing data as well as on query and display. A SDBMS provides geo-query and thematic mapping of existing data but no spatial analysis. Mapping and graphics packages provide colorful map products but no query or analysis capabilities. The differences in the capabilities among these packages arise from their differences in topological structures. Topology differences also account for important differences in cost and complexity of precision farming software.

Use of Metadata in Precision Farming

Metadata refers to "data about data," including specification of the spatial and thematic characteristics of mapped data. Five elements make up the metadata used in precision farming:

— data dictionary (measurement units, relationships),

— thematic data attributes (integer, floating-point),

— spatial data structure (vector, raster),

— geo-referencing (coordinate system, projection) and

— data quality (derivation procedures, level of detail).

The *data dictionary* might define "bushels per acre" as the common unit of measurement for wheat and "pounds per acre" or "kilograms per hectare" as interchangeable definitions for nitrogen applications. Once the units are identified, the thematic data attributes for each data element are defined and the structure of the thematic table is specified. For example, yield might be characterized as a positive, floating-point value while crop type might be set to a positive integer code and a large alphanumeric text string.

At this point the computer knows about the "what" information for each map variable. The spatial data structure identifies the composition and organization of the spatial table. Geo-referencing details the procedures used in registering the map. Now the computer knows "where" the data are on the face of the earth. Finally, data quality specifications complete the description of the data by noting "how good it is."

A benefit of adhering to common metadata definitions is the interchange of data among different software. Early systems were self-contained and relied on nonstandard (often proprietary) data formats. It was like the old days when a red hitch wouldn't accept a green implement. The details encapsulated in the metadata describe a data set in terms of its format and structure, which is particularly useful for software developers. But it also deals with interrelationships and common descriptions. It organizes data in meaningful terms for intended users who want to "surf" the conceptual framework of related data.

The details of metadata best are left to programmers, but the agriculture community needs to understand the importance of standards in the products they buy. The first four elements of metadata facilitate data exchange by detailing the "what" and "where" information while the data quality element describes its accuracy. Statistics on the level of detail and data confidence should be the focal point of end users (farmers)—"garbage in is garbage out," no matter how efficiently it is transferred.

We have identified the distinction between precision and accuracy and even suggested techniques for assessing and mapping uncertainty. Several areas of development in GIS technology extend these measures, thereby improving the data quality component of metadata. A map pedigree follows the lineage of a map like a hog's pedigree follows its genealogy. It is similar to the textual descriptions accompanying soil maps, with some significant enhancements.

A *map pedigree* logs in information such as who is "responsible" for the data, how base data are collected, how derived data are processed and what are the statistics on map precision and accuracy. For example, a GIS model of soil loss might use a base map of field elevation to calculate a derived layer of terrain slope. In turn, the slope map is interpreted for its effect on soil loss (increasing slope increases soil loss). The interpreted slope map is combined with base, derived and interpreted maps showing other factors (e.g., soil type, crop cover, tillage) to generate a modeled map of soil loss for a series of storm events. A map of crop production potential might be generated from similar set of mapped data. Or a prescription map for site-specific nutrient application might be modeled from a set of maps depicting the current levels of soil nutrients throughout a field.

What is common among these examples is that the modeled map is dependent upon logic and processing flow and not "the-precise-placement-of-physical-features-for-navigation" heritage of the traditional map. The GIS technology has extended the descriptive role of mapping to interpretation, prediction and prescription. As such, the concepts of map precision and accuracy must be extended to model logic and interactions among mapped data.

A *logic flowchart* is a diagram illustrating the interrelationships among maps leading to a modeled map. It is a graphical summary of the map pedigree for maps that interpret, predict or prescribe spatial conditions. It shows you each "thoughtful" step in the process with boxes representing maps and arrows connecting the boxes representing processing. It extends traditional data quality measures to the quality of the "conceptual" maps used in decision making.

5

Affordable Opportunities for Precision Farming

Typically, precision farming is associated with sophisticated computer and satellite technologies, which some farmers have neither the physical nor the financial resources to implement. However, even without expensive technologies, precision farming may be used to enhance a wide variety of crop management decisions. A few examples are provided here, but there are others. Any farming practice that can be managed spatially (i.e., across different geographic areas) or temporally (over time) can be improved with precision farming. Examples of management that may benefit from precision farming follow:

Soil fertility and soil quality: Yield levels within a field may be correlated with soil nutrient levels. Decisions regarding fertiliser rates, where to apply lime, and where and how much manure to apply may be based on this information. Information about topsoil depth, organic matter, and traffic patterns can be used to help manage compaction, drainage, water holding capacity, erosion, and buffer strips.

Pest management: Scouting for pests can be used in sections of fields in the same way it has been applied to whole fields. Insects, diseases, and weed infestations tend to be clustered together in patches. Basic information technologies can help determine where these infestations occur or are most likely to occur. Farmers may then make decisions regarding site-specific treatment that can save money and reduce the risk of water contamination from a widespread application of pesticides.

Planting decisions: Variety choice, seeding rate, and seeding date decisions may benefit from site-specific information. On sandy soil or south-facing slopes, drought-tolerant varieties may be needed. On saline soils, salt-tolerant plants may be needed. In cooler areas, such as on north-facing slopes, seeding dates may be delayed and seeding rates may be increased to maximise yield potential and minimise the potential for crop failure. Enterprise Placement. Enterprises must be located only on suitable sites. Knowing the spatial characteristics of the farm can help farmers determine where individual enterprises are most appropriate.

Costs and Benefits of Precision Farming

Precision agriculture is usually associated with sophisticated technologies that vary in price. The common technologies associated with precision farming include computer hardware and software, GPS and yield monitoring equipment, and equipment for variable rate applications. The cost for these technologies may range from several thousands of dollars to tens-of-thousands of dollars, plus $10+ per acre for an initial soil grid soil sampling with a 2.5-acre grid size. Annual fees for weather information, variable-rate application, and related services may cost as much as $12 per acre.

Despite these costs, precision farming can sometimes pay off if it results in more efficient use of resources and inputs. When a whole field is treated uniformly, some areas will receive excessive amounts of fertiliser or pesticides that will not result in higher yields. Other areas will not receive enough inputs. Or perhaps spot treating a pest infestation or deep ripping a small portion of a field will generate a greater yield and income benefit than if the same amount of money were put into a different input spread across the whole field.

The goal of precision agriculture is to gather site-specific information and enable farmers to vary management practices more appropriately within fields. These principles of precision agriculture are relevant to small and large farms and to high-tech and low-tech approaches to gathering information.

Spatial Variability

No field is uniform. Soils, historic use, microclimate, and pest pressures can vary significantly over short distances. Variability can be caused by geologic, topographic, atmospheric, and human factors. Knowing sources and kinds of variability can help farmers recognise these differences and manage for them.

As landscapes were shaped over geological time, spatial variability emerged as a natural part of soil formation (fig. 1). Parent materials (the rocks, minerals, and organic material from which soils developed) differed from place to place and from time to time; erosion and deposition rearranged the depth of soil across the landscape. Lowlying areas gained topsoil and nutrients from surrounding land; higher ground may have been stripped of nutrients through erosion and leaching.

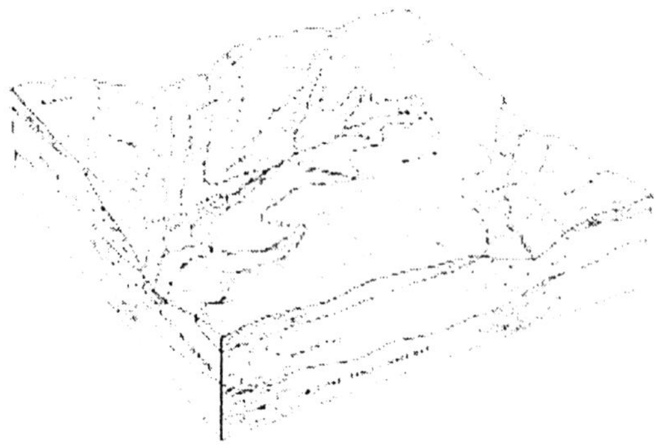

Figure 1. Differences in geologic parent materials and landscape position cause soils with different characteristics to form over time.

Changes in agricultural land use have also created spatial variability. Animal enclosures-a frequent part of farming before tractor mechanisation-created areas of nutrient deposition. Many wood lots, pastures, windbreaks, and fence lines, common at the turn of the twentieth century, have been removed, leaving variations in the soil resource. In a few years, or even within a year, weeds and pests can create their own spatial variability and problems. The dispersal of seed or the spreading of perennial weeds by underground shoots can lead to clumps of weeds near invading plants. Windborne insects and pathogens may infest a crop in a random pattern. Some areas of a field may be severely damaged, while other areas remain unaffected.

Farmers also cause spatial variability. Tractor tires can create compacted areas, especially in turn-rows. Inconsistent maintenance of spray nozzles and injector knives can cause row-to-row variability in crop vigour due to the differences in application rates. Broadcast amendments often have strips with double the application rate, where the application path overlaps, or no amendments at all, where the applicator misses the target. So, nothing in the real world of agriculture is uniform.

Scale of Variability

The scale of variability is the distance over which conditions change. For example, patches of weeds may cover an acre, while nutrient levels may differ over distances of inches under a history of banded fertiliser applications. The scale of variability determines the opportunities to manage the variation. If conditions vary little within fields (i.e., the scale of variation is quite large), fields can be sampled and managed as single units. No new technology is needed. Precision farming technology is also not needed if the scale of variation is quite small because it is not practical to manage small sites differently. Each problem has its own scale of variability so different management zones should be developed for each problem. A flood-prone section may form a single large zone in the lower part of a field, but spots of sandy, eroded, compacted, or high pH soils may need fine-scale management targeted to areas of less than an acre.

The value of site-specific management, both in determining and solving problems, depends on the nature of the spatial variability. When assessing whether spatial variability justifies the investment of time and money required to implement precision farming, it is helpful to break the problem into some separate questions:

1. *What are the extremes?* If all of the soil test values are above the critical threshold for adding fertiliser, then it does not matter how much spatial variation there is - no fertiliser should be applied to the field.

2. *Is the variability observed in large patches?* If it is not observed in large patches, then it may be impractical to vary management over the small distances that the changes occur.

3. *How skewed (lopsided) is the variability?* Most soil properties are skewed. That is, very few places within a field match the average measured for the field, and some places may be much higher or lower than the average. When a field is highly skewed, managing for the average situation is most likely to have poor results because most of the field will not be close to the average (fig.2). Implementing precision farming can be cost effective in these situations.

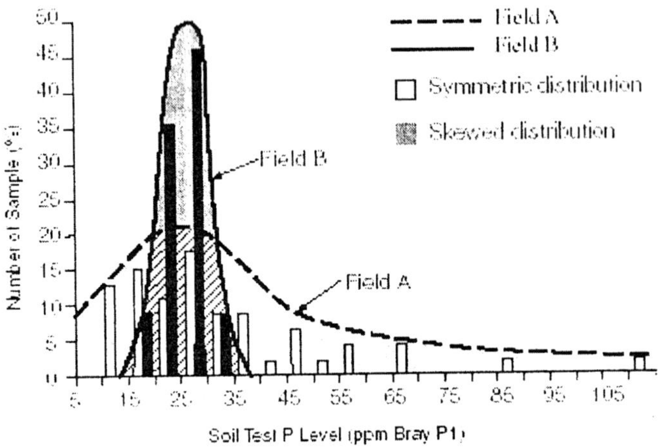

Figure 2. Results of a study in which 45 soil samples from each of 2 fields were analysed for available phosphorus.

4. *Does crop productivity vary in a gradual manner from one part of the field to another, or are there*

sharp boundaries of differing yields? For example, does yield increase (or decrease) gradually from the top to the bottom of a slope, or does it abruptly increase or decrease at some point? Yield patterns may give clues about the reasons they occur and may be helpful in establishing management zones.

Low Cost Record Keeping Method

By combining existing maps with maps of farmer observations, measurements, and management practices, farmers can learn about patterns and causes of variability on their land. The goal of precision farming technology is to improve the efficiency of gathering measurements, combine them into informative maps, and vary management practices based on the more detailed information. This section describes a low-cost spatial record keeping procedure that does not require a computer.

Materials needed for this procedure include:

— County soil survey provides soil maps and land use interpretations of the farm to be mapped. Soil surveys may be obtained from the local county Natural Resources Conservation Service (NRCS) office, Cooperative Extension Service, Soil and Water Conservation District office, or possibly a local library. Information available in the soil survey includes soil descriptions, land use suitability and/or limitations, and types and locations of other soils that may be in the vicinity.

— Aerial photograph of the farm may be obtained through the local NRCS or Farm Services Agency (FSA) offices, or may be downloaded from the Internet.

— Clear transparency sheets or other see-through, standard letter-sized sheets are available at office or art supply stores.

— Permanent transparency markers or other felt-tip markers are generally available wherever school supplies are sold.

— Farm data and records, such as soil testing information, precipitation and other weather data, historical land use records, and recent management practices.

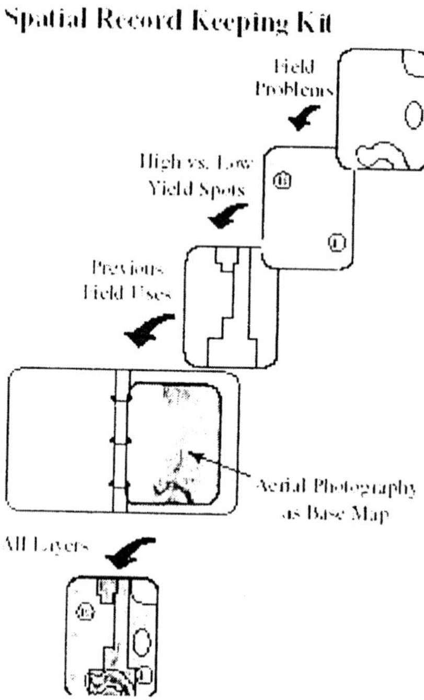

Figure 3. The spatial record keeping kit can help the farmer keep track of on-farm variability and how it changes over space and time.

— A three-ring binder or other book to organise and protect the spatial record keeping kit (fig. 3).

Step 1. Obtain a base map

This map will be the background for all the transparent overlays. See the section below for suggestions of sources of maps. A soil survey map is recommended because it contains soil mapping unit lines superimposed on an aerial photograph. Photocopy the pages of the soil survey that contain the farm. Enlarging the soil map on a copy machine makes it easier to view. If an aerial photograph or digital orthophotograph is used as the base map, record soils information on a separate overlay transparency.

Step 2. Create an outline map

This map will be a simple line drawing of the boundaries and major landmarks on the farm. Lay a transparent sheet of plastic over the base map and use a permanent marker to trace the boundaries of fields, roads, streams, farmsteads, and other landscape features to use as reference points. By making copies of this on paper or plastic transparencies, the outline can be used as a base for recording other information.

Step 3. Record soil information

Whether a soils map is used as a base or an overlay, read and record notes about the characteristics of the soil mapping units on the farm. Mapping units may be designated with numbers or a combination of letters and numbers. The soil survey provides a detailed description of each mapping unit. Record information from the soil survey about the soil texture of the A and B horizons, slope class, and management characteristics, such as drainage class, stoniness, and water holding capacity. The soil survey also has useful interpretive ratings for water management, irrigation potential, and other land uses.

Traditional soil surveys are called Order 2 surveys, meaning that each mapping unit is actually a mix of different soils. If yield varies significantly across a field, but the soil survey indicates that only one type of soil is present, the variation may be due to soil inclusions. A soil scientist may be able to explain at least some of the differences observed.

Step 4. Create overlay maps

Draw a series of overlay maps that can be stacked to identify similarities in patterns. Each transparent overlay should show only one type of information and should be labeled with the topic, date, and time of observation. They must be drawn on the same outline map or show the same landmarks so they can be lined up and compared. Use a different pen colour or line pattern for each sheet so each piece of information can be differentiated when several transparencies are stacked together.

Maps of some information, such as land use history and topography, need only be drawn once. Other information, such as weed patterns or ease of tillage, can be drawn at regular intervals to track changes in these properties. Overlays may show information, such as:

- — Topography (high spots, low spots, and/or steep areas)
- — Weeds (type of weeds present and size of patches)
- — Yields (areas of noticeably higher or lower yields, or actual yield measurements)
- — Crop growth characteristics (leaf colour, stunting, disease)

— Insect damage and type of insect
— Soil moisture observations (droughtiness, ponding after rain)
— Ease of tillage
— Soil colour
— Drainage patterns
— Erosion
— Earthworm counts (biological life and activity)
— Compaction problems
— Presence of rocks
— Previous management practices (crop rotations, tillage)
— Precipitation rates and weather conditions
— Crop prices (or other economic information)
— Land use history (location of old feedlots, fence lines, etc.)
— Management records (planting date, seed variety, pesticide and fertiliser treatments)
— Soil test results (pH, nitrogen, phosphorus, potassium, organic matter)

Interpreting the Spatial Data

After the observation maps are drawn, the next step is to look for relationships among the landscape features. For example, hard-to-work soils may be consistently associated with low-lying areas, droughty areas may be associated with particular pest pressures, or areas with consistently low yields may appear as wet areas. High phosphorus levels may correlate with the location of an old feedlot. Insect damage may relate to slope. Overlaying the individual transparencies should help identify some patterns. Maps from multiple years of observations may help confirm or refute observed and suspected relationships based on years of farming the land.

A further step is to identify management zones and record the results of management decisions on new transparencies. Management zones are areas within a field that are managed differently from other parts of the field. The boundaries of management zones are determined by variations in land characteristics identified on the maps. As described in the section above about scale, a management zone is only useful if the farmer is able to vary management at the scale of variation of the land property.

Farm equipment sets practical constraints on the size and shape of management zones. Equipment that requires a significant amount of time to change rates is impractical for fields divided into many management zones. Variablerate controllers are designed to change rates on-the-go, making fine-scale management zones practical. Smaller equipment also helps in managing at a finer resolution. Eventually, it may become clear that more specific quantitative information would be worthwhile to gather. The hand-drawn maps can be used to plan application of a geographic information system (GIS), global positioning system (GPS), and/or other site-specific management technologies. Hand drawn maps are a low cost form of GIS.

Using Interpretations to Make Decisions

Documentation of spatial variability is not particularly useful if it cannot help farmers make better management decisions. Interpreting the maps can help answer questions, such as: Is a significantly lower yield in one part of the field a result of poor drainage, low organic matter, pest pressures, or other factors? Is the variability manageable? Is it cost-effective? The more spatial information available, the more accurate and useful interpretations may be. Since climate interacts with soil and other factors to affect yield response, several years' records of spatial information may be needed to see the full benefit of the interpretation. After identifying soil variation and possible causes, there are two ways to respond: reduce the amount of variation by selectively applying amendments (such as lime to even out the pH across a field) or adapt to the variation (such as by planting different varieties on different parts of the field).

Collecting Spatial Information

Sources of Maps

Soil surveys are available for most farmland in the United States. Soil surveys provide spatial information about soil properties, such as texture, drainage, seasonal wetness, depth to bedrock, erosivity, and slope. Aerial photographs are available for most farmland from a variety of sources including county and state planning offices and the local NRCS and FSA offices. The federal photographs are printed at a scale of 8 inches equals 1 mile and are appropriate for field-scale planning. Each photograph usually covers 1 square mile of land. Some photographs available through NRCS may be more than 10 years old, but they still provide valuable spatial information about the farm. Older photographs can also provide excellent historical information. FSA maintains current aerial photographs of farmland as photographic slides, which may be copied at a nominal cost. The slides, however, need to be enlarged at the user's expense.

The enlargement changes the original scale, so a new scale needs to be calculated. This can be done by measuring the distance between two known landmarks (such as the farmstead and barn) on the photograph and comparing this distance with the measured distance between them on the ground. It is important to know that scale is not accurate on aerial photographs unless they are orthophotographs; that is, they have been corrected for the distortions caused by camera tilt, curvature of the lens, and topography of the ground. As relief increases, the need for orthophotographs increases.

Digital orthophotographs are orthophotographs that have been digitised into electronic format and corrected for distortions so that distances on the photograph are directly proportional to distances on the ground. Digital orthophotographs can be aligned with other types of digital data to create GIS layers for the farm.

Infrared photography is a special kind of photography that is sensitive to light wavelengths that are reflected from living organisms, but invisible to the eye. Infrared photography can provide information on the health and vigour of growing crops that standard aerial photography cannot provide. Infrared photographs of local farms are available in a few NRCS offices.

Topographic maps are widely available pieces of information that show the lay of the land. The U.S. Geological Survey (USGS) produces these maps at a variety of scales, and they cover the entire United States. USGS maps of local areas are often sold in bookstores or may be ordered directly from the USGS.

Climate data may be helpful in interpreting variability observed on the farm. The NRCS National Water and Climate Center, in cooperation with Oregon State University, has developed detailed maps of average annual precipitation that may be obtained from each state NRCS office. These electronic maps require the use of a

GIS software package, such as ArcView, ArcInfo or ArcExplorer.

Some of the most valuable spatial information is obtained by direct observation, personal experience, or oral history. Farmers get to know the land they farm over time. Given the opportunity, they usually pinpoint areas of special concern. These areas may be poorly drained or droughty, consistently produce higher or lower yields than the rest of the farm, warm more slowly in the spring, or be frequently prone to heavy pest pressures.

To manage for site-specific differences, it is helpful to record these types of observations continuously along with the specific information used to identify areas of concern (for example, crop yields or soil tests). Soil Quality Health Cards are excellent for directing a farmer through detailed observations of approximately 10 crop and soil parameters. Over time, observations can provide a fairly clear picture of the spatial variability that may respond to management.

Site-specific Yield Estimation

Monitoring Differences

One of the most effective ways of evaluating variability is by monitoring differences in yield across a field. Yield differences usually reflect more basic differences in soils, topography, microclimate, or weed competition and can give farmers a good feel for those areas that might respond well to site-specific management. A popular tool of precision farming is the computerised yield monitor, a device attached to a combine that estimates yield by measuring the flow rate of the crop being harvested and the area covered by the harvester.

When attached to a Global Positioning System (GPS) unit, farmers can map out the variation in yield across a field. Yield monitors are not the only way to learn how production varies inside fields. With a little care and not much expense, farmers can spot check yields to identify the most productive areas and those needing extra attention. These methods are valid for measuring relative yield-identifying the best areas and the worst areas. They are less useful for estimating actual yield values. Table 1 lists some techniques for estimating yield without yield monitors.

These measures can be recorded on transparencies and added to the spatial record keeping kit. Farmers who check site-specific yields soon realise that looks can be deceiving. Grain yield does not always match how quickly the crop got started in the spring or how tall, dense, or green the crop looks at the end of the season.

Direct count. Some yield estimates, especially for cereal grains, are based on simple counts of the things that make up yield: plants, pods or ears, and seeds per pod or ear. To estimate corn yields, first count the number of ears in a measured area (for example, a 6-foot row length in a field with 30-inch rows). Then, count the number of rows of kernels on representative ears (possibly every third or fourth ear) and the number of kernels per row on representative rows. Multiply the numbers (number of ears/field area, rows of kernels/ear, and kernels/row of kernels) to get kernels in the field area. Assume yield is proportional to that number.

Kernels for other cereals are more difficult to count. In fields with good stands, the number of spikelets per head of wheat can be used as an indicator of yield potential. Soybeans are more complicated than the cereals. Pod numbers do not correlate well with yield, so counting pods per plant or in an area is not ideal. Soybean yields are sensitive to the number of seeds per pod and the weight of the filled seed, aspects not captured in pod counts. Instead, weigh the whole plant or use a weigh wagon.

Weights. In some cases estimating yields directly is simpler than relying on counts. Hand-harvest a standard area, bundle the sample, hang it in a barn to dry, and weigh the air-dry samples with a scale. Assume yield is proportional to the weights. Ear samples are convenient for corn. Whole soybean plants and the heads of sorghum and small grains can be clipped with hand shears. Hay can also be harvested in small plots by clipping with hand shears. These observations do not have to interfere with the labour crunch of harvest time.

Sampling can be done before the grain is dry enough to harvest, and the weights can be measured in winter after the harvest is finished. Interpretations. No single method for estimating sitespecific yields works in every case. There are always trade-offs. Faster methods may be less precise, but more locations can be checked and observations are more likely to fit into a farmer's work schedule. Whatever method is selected, make it systematic.

Carefully follow the same procedures at every site so that the estimates of sitespecific yields are comparable throughout the field. Use caution in interpreting yield information. Relative yields are not always consistent across a field from year to year. Yields may be higher than average in one section of the field one year and lower than average the next year. The factors affecting these differences are not always understood, but weather plays a role. Major management decisions should not be based solely on one or two years' yield data. Yield differences are indicative of underlying differences in the site. Determining what those differences are is the challenge the farmer faces. Knowing the cause of the yield variation helps the farmer make decisions regarding what to do about it.

Soil Sampling Procedure

Spatial variability on a farm can be determined and managed by directly observing crop responses in different places on the landscape. Sometimes variability does not seem to relate to landscape position or general soil characteristics. In these cases, a detailed soil sampling procedure may be worth the investment. Grid soil sampling is a procedure used to obtain an unbiased assessment of the variability that exists in a field. It involves laying a grid over a field and testing soil samples at the corners or center of each grid cell. The grid spacing may be uniform or may have a range of sizes. The results of the grid soil sampling can be used to create a detailed soil fertility map or series of maps that indicate the variability of measured soil properties throughout the field. Grid soil sampling is fairly expensive, and unless the grids are fine enough, the resulting soil fertility maps may be misleading.

A less expensive method of soil sampling for spatial variability is directed soil sampling. Directed soil sampling focuses individual samples in areas that are likely to have different characteristics. For example, areas in which yield is consistently higher or lower than the rest of the field, areas with minor landscape fluctuations, and areas with distinctly different soil types may warrant separate soil samples. Over time, these samples could help the farmer establish management zones within the field.

Each management zone must be tested or monitored separately. When sampling an area, producers must be aware of variation within the area they are testing to avoid systematic bias. For example, always sampling in the bottom of the furrow can lead to a result different from sampling the entire area between rows. Farmers avoid this problem by collecting soil from a variety of positions and mixing it well before subsampling and sending it for analysis. Soil pH, organic matter, and phosphorus often vary by depth, so the soil must be sampled to exactly the desired depth to avoid misleading results.

Soil fertility maps correlate with yield to help determine management strategies within a field. However, soil fertility maps do not account for the overriding effects of weather on yield, and may be misleading if these effects are not recognised. These issues can be addressed by keeping records over several seasons (temporal variability record keeping). What is most important to remember is that the more farmers know about the variability on the lands they farm, the better the management decisions they can make.

Soil Quality Assessments

Soil quality is how well soil performs functions, such as crop production, water retention, or water filtering. It can be assessed using several on-farm techniques including the Soil Quality Test Kit and the Soil Health Card.

On-farm Testing

On-farm testing is an excellent tool for helping farmers determine sources of variation and appropriate management strategies for addressing those sources of variation.

6

Precision Agriculture and Environmental Quality

The advantages of PF is it offers opportunities to improve agriculture productivity and product quality, reduces agro-chemical wastage through efficient application and resulting in minimizing environmental pollution and in energy conservation. In PF timeliness of in-field operations (cultivation, seed sowing, application of fertilizers & pesticides and harvest) is crucial.

PF has, therefore, not only the ability to apply treatments that are varied at local level, but also to precisely monitor and assess the agricultural enterprise at a local and farm level. It also provide sufficient understanding of the processes involved to apply inputs in such a way as to be able to achieve a particular goal. The goal, however, might not necessarily mean maximum yield but may be to optimise financial advantage while operating within environmental constraints.

Agricultural practices and their effect on the movement of sediment and agricultural chemicals have been a source of controversy because of the inability to define precisely the impacts of various farming practices on nonpointsource pollution. Nonpoint-source pollution links agricultural practices with offsite movement of nutrients, pesticides, and sediment. The exact relationship, however, depends upon soil, topography, meteorology, cropping practice, and landscape. What role agricultural practices have on the amount of movement of these constituents from the edge of fields has been debated for years, with little agreement about the efficacy of various practices on environmental quality. But there is agreement about the fact that agricultural practices that either reduce inputs or apply inputs at times when there is less risk of offsite movement have a positive impact on environmental quality.

Several emerging issues regarding nonpoint-source pollution should be considered in any discussion about agriculture and the environment. These include the movement of pathogens and antibiotics into nearby water bodies from agricultural fields that have had manure applied to those fields. The discussion of offsite movement is complicated, however, by one important fact: while water is the primary transport mechanism, the processes that enable the movement of nutrients, pesticides, pathogens, and sediments differ and, therefore, respond differently to management practices.

It is also important to remember that environmental quality is multi-dimensional. A number of interrelated factors are linked to offsite movement. These include soil management practices, soil type, topography, organic matter content, crop, weather events, and prior management. In many cases there is movement within a field but no loss from the field. This redistribution of sediment, nutrients, residue, or pesticides may be indicative of a suite of management practices that is less than optimal in terms of performance. Precision agriculture or site-specific agricultural practices may affect offsite movement of different components from a field. There have been few studies conducted with the specific objective of evaluating the potential role of precision agriculture on environmental quality.

Precision agriculture or site-specific management refers to the differential application of inputs to cropping systems or tillage operations across a management unit (field). Input applications may vary either spatially or temporally within management units. The methods involved include application via predefined maps based on soil or crop condition or sensors that control application as machinery traverses the field. Spatial scales range from less than 1 meter to more than 10 meters, while temporal scales vary from minutes to months.

Environmental quality endpoints require definition because of the need to understand how the performance of management practices is to be measured and compared among practices. Nitrate-nitrogen in drinking water has a Maximum Contaminant Limit (MCL) of 10 milligrams per liter, while atrazine has a MCL of 3 micrograms per liter. These two agricultural chemicals are present in agricultural systems, and the MCLs are used to determine if environmental quality has been compromised as a result of farming operations. Rigorous measures of environmental quality are needed to assess the impact of various management practices.

A farming system is comprised of many elements. Importantly, three classes of variation exist within a field: (1) natural, such as soil and topography; (2) random, such as rainfall; and (3) managed, the fertiliser or seed application. Offsite impacts are the result of the interactions of these three sources of variation. Another aspect that must be defined is the interaction between natural cycles within soils and supplied inputs. An example is the interaction between mineralisation of nitrogen from organic matter, a temperature-dependent process, and the application of commercial sources of nitrogen at different times of the year.

Each of these supply nitrogen to the soil profile, but the nutrient's availability relative to crop uptake patterns differs between the cycles. These differences often lead to some of the environmental impacts from agricultural systems. Nitrogen available in the early growing season provides more than is needed to meet crop demand; if precipitation then exceeds crop water use, substantial nitrogen leaching can occur. These patterns must be understood to place precision farming and environmental quality in the correct perspective.

Soil Variation

Soil variation is a spatial variable. Water-holding capacity or organic matter variation, along with topography, provides even a more interesting view of a field in which a producer places inputs or disturbs the soil. Other variables could be layered within this field to create a series of interacting elements. The central question is how to quantify soil variation. Wollenhaupt and colleagues provided a detailed summary of soil sampling and interpolation techniques that potentially could be used to quantify soil variation. They defined these sampling methods as judgmental sampling, simple random sampling, stratified sampling, cluster sampling, nested or multistage sampling, systematic sampling, stratified systematic unaligned sampling, and search sampling.

Selection of a sampling method depends upon the goal of sampling. Collection and analysis of the samples only provides one portion of the base layer of information. Individual samples represent points; to be of value, they must be interpolated. There are as many interpolation schemes as there are sampling schemes. These schemes include central tendency, proximal, inverse distance methods, splines, and geostatistical. None of these methods or schemes are described herein, but all have been used to determine variability across a field.

Topographic variation within fields can be collected from topographic maps, but the resolution on these maps is often insufficient to provide the necessary detail about variations within fields. Topographic maps can be generated from differential or kinematic geographic positioning systems. The role of topographic variations on water use, plant growth, soil processes, yield, surface runoff, and groundwater hydrology has not been quantified for agricultural fields.

Biological Variations

Biological variations within fields are as great as soil variations. A number of measures of biological variation must be considered. These include soil microbial populations, weed populations, insect populations, disease occurrence, crop growth, and harvestable yield. The most commonly observed variation is through yield maps generated by producers with yield monitors. Growth and yield differences across a field can be large; the magnitude and often surprise observers. Jaynes and Colvin determined that yield variation was due to an interaction of soil types and precipitation patterns within the growing season. In their data set, which consisted of 7 years of observation across the same field, there were large differences in yield.

Insects, diseases, and weeds are variable factors within fields. But the processes leading to variation within fields are quite different. Insect patterns within fields vary as a result of migration into the field. Fleischer and colleagues examined spatial variation in insect populations and identified several interacting that lead to spatial variation on a field scale. Their conclusion:

Within a field, migration, colonisation, and reproduction are processes that lead to increases in population, while emigration and mortality reduce the population. But there are a number of interacting factors, for example, temperature, stage of crop development, and host plants, that will affect insect behaviour and development. Comprehensive models of insect growth and populations spatially across a field have only begun to emerge. Insect patterns across a field also can have a large temporal component because these population factors are often temperature-dependent.

Spatial distribution of disease outbreaks within fields has been observed for a number of years. Observations of spatial occurrence of diseases have led to the development of integrated pest management programs for a number of crops.

Weed distributions within fields have been a topic of interest to both producers and weed scientists. Andreasen and colleagues evaluated different soil properties to show why populations of 37 different weed species varied across fields in Denmark. In the case of weeds, as contrasted to insects or diseases, the seedbank dynamics across a field become the primary factor affecting the spatial distribution of weed populations. Hatfield showed how microclimatic conditions at the soil surface, as affected by crop stand and soil characteristics, could play a role in the establishment of weed populations across a field.

This model describes the interactions between radiation levels at the soil surface and the energy balance in determining the likelihood of a weed seed germinating and surviving. Johnson and associates used geostatistical methods to estimate the spatial and temporal variations in weed seedling populations; they found that this methodology could be effectively used for these analyses. They found also that the location of velvetleaf remained constant across years but the density of weeds within an area varied among years. Earlier, Johnson and colleagues had found that populations depended upon prior management practices within a field.

Spatial variation of weeds within a field depends upon seedbank dynamics. Temporal variation patterns reflect interactions among soil, meteorological conditions, crop canopy, and growth characteristics of the specific weed. If these factors are layered together, a picture of the potential interactions that influence spatial and temporal weed patterns across a field begin to emerge.

Soil Process Variation

Nitrogen in the soil profile represents the balance that exists among a number of processes. These processes include mineralisation, immobilisation, denitrification, volatilisation, nitrification, sorption, plant uptake, and leaching. Mulla and Schepers detailed the properties that are related to these processes; many are associated with soil water content, soil temperature, soil pH, soil texture, organic matter content, and soil drainage status. Nitrogen is one of the critical elements for plant growth. Processes that determine nitrogen availability within the soil-mineralisation and nitrification-and processes that make nitrogen unavailable-denitrification, volatilisation, and immobilisation-vary with soil types. It is reasonable to assume that inherent changes in soil would have a direct effect on plant growth and yield. The challenge is to quantify the response to varying nitrogen levels.

Nitrogen is soluble in water, and it moves rapidly in the soil profile with water. This is particularly evident in subsurface-drained soils where nitrate-nitrogen concentrations often exceed 15 milligrams per liter. Hatfield and associates showed that nitrate loads from drained fields were linearly related to precipitation totals. This coupled with the variation in soil water use patterns across a field can give rise to variations in the amount of subsurface drainage throughout a field and within and across years. The complexity of the interactions between the physical environment and the biological response creates a situation in which it is difficult to quantify the response to different practices.

Environmental Problems

The concerns can be categorised as endpoints in different water bodies, for example, surface water and groundwater, by a number of different indicators, such as nutrients, pesticides, sediment, manure, feed additives, pathogens, and heavy metals. The accompanying table summarises where the environmental risks are perceived to be the greatest for different processes.

The interactions between factors and processes must be addressed in any discussion of environmental quality. Nitrate-N losses are influenced by any factor that affects the movement of water within and from the field. Because nitrate-N is soluble in water, it is readily transported with any moving water. Sources of nutrients in the upper Midwest have been catalogued by Goolsby and colleagues in their examinations of agriculture's contribution to the hypoxic zone in the Gulf of Mexico.

Hypoxia is a condition where water is depleted of its oxygen content, which results in a serious reduction of biological activity.

Environmental risks from nutrients and soil organic matter

Process	*N*	*P*	*K*	*S*	*OM*
Leaching	+	0	-	-	-
Denitrification	+	-	-	-	-
Eutrophication	+	+	-	-	-
Precipitation	+	+	+	-	-
Runoff	+	+	-	-	+
Volatilisation	+	-	-	0	-
Saltation	-	-	+	-	-

An area-approximately 7,000 square miles-in the Gulf of Mexico where the Mississippi River empties has developed into a hypoxic zone, with very low fish and shellfish densities. The hypoxic zone is believed to be caused by an abundance of nitrogen, phosphorus, and silicate in the water. The table of environmental risks provides a basis for examining agriculture's impacts on environmental quality and the magnitude of those impacts. Schepers suggested that hypoxia could result from runoff, erosion, subsurface drainage, and baseflow from fields into adjacent streams. These transport processes, when coupled with nutrients, heavy metals, pesticides, antibiotics, pathogens, and sediment, give rise to differential movement of these contaminants across a watershed.

The processes outlined in the table cannot be changed, but it is possible to modify the loading of nutrients and pesticides in a field. Nutrients can be applied as inorganic sources (commercial fertiliser) or organic sources (manure or compost) to a field, and the timing of applications may be an effective method of avoiding excessive nutrients or pesticides when transport mechanisms are favourable for off-site movement. Nutrient application requires an understanding of crop requirements for profitable yield, efficiency of crop utilisation, ability of soil to supply a portion of the crop requirements, and temporal patterns of crop uptake and utilisation compared to supply from the soil volume. Application of nutrients provides an opportunity for effective management of inputs while increasing production efficiency. Two aspects of nutrient management become critical:

1. What is the soil supply of nutrients, and how can this be measured or predicted over a diverse landscape?
2. How can the plant be used as an indicator of nutrient status?

Both questions require attention to better understand nutrient management and its use in precision agriculture for environmental quality improvement purposes.

Pest Management

Pests associated with agricultural production include weeds, insects, and diseases. Most pesticides applied in agriculture are for weed control. The environmental problems facing agriculture from pesticide use have to do mainly with groundwater and surface water quality. Dave Mortenson explained to workshop participants, from the perspective of weed ecology within a field, that weed populations vary in response to a number of factors and that understanding the dynamics of weeds is critical to developing production systems that reduce the amount of herbicide applied to the soil or crop. Weeds are spatially variable across fields because of organic matter, soil texture, landscape position, and the interaction of these factors with crop management, crop cultivars, tillage, planting density, cultivation, and herbicide application methods. This is a complex set of interactions that exists within all fields.

Precision agriculture provides an enabling set of technologies to help reduce potential environmental problems from pest management. These technologies include field maps of weed distribution throughout the year, detection methods for weeds within a field, application methods to apply an herbicide only on selected areas, and indications of the effects of weed populations on crop yields with data from yield monitors on combines. Each of these technologies allows for the adoption of precision agriculture to weed management. Insects and diseases can be treated similarly to weeds using the same principles.

Soil and Water Quality for Sustainable Agriculture

Soil and water quality are two major components of a sustainable agricultural system. Attributes of soil and water quality are inextricably linked. A good soil does not ensure good water quality, but a poor soil is likely to create conditions that contribute to poor water quality. Soil quality must be maintained to provide a supply of nutrients and water and support the developing crop. A healthy soil is thus critical to food, feed, and fiber production. Water quality problems are often considered to be pesticide- or nutrient-based; however, the largest water quality problem results from sediment due to soil erosion. Soil and water quality are linked because erosion processes remove valuable topsoil and transport it offsite, where it often becomes a water quality problem.

Soil quality can be defined as fitness for use. Soil quality is the capacity of a soil to function in a productive and sustained manner, while maintaining or improving the resource base, environment, and plant, animal, and human health. The capability of a soil to function within ecosystem boundaries and interact with the environment, external to that system forms the basis for determining the potential impact of soil management systems on the environment.

Soil quality measurements are needed to guide development of management practices. The land manager needs quantitative information about whether a given management practice is degrading, aggrading, or maintaining the soil's ability to serve its intended use. Further, the manager needs to know what the status of each major attribute of the soil is so changes can be made. Measurement of soil quality should be a quantitative tool to help guide management decisions.

Because soil is such a complex material and because it varies temporally and spatially, Larson and Pierce suggested that a "minimum data set" of important parameters are needed for each soil from which measures of quality could be computed. Doran and associates suggested a similar approach. Because many soil attributes are interrelated, only a limited number of attribute measurements are needed. The measured attributes can be used to estimate other attributes by equations called pedotransfer functions. For example, if the amount and type of clay, organic matter content, and bulk density are known, the soil's water-holding capacity can be estimated. The measured or estimated attributes can then be used with suitable models for any intended use of the soil. Doran and colleagues pointed out that while an analytical methodology is needed for researchers' in-depth studies a more easily determined and recognisable set of indicators is needed for land managers' use.

The variability of soil nutrients within a management unit or field has long been known, as has the susceptibility to erosion. Recent yield monitoring suggests many other factors affecting soil behaviour are extremely variable. Soil compaction and the resulting water drainage appear to be more common than previously thought. The discovery and quantification of these causes of variability suggest many new applications of precision agriculture and the need to develop new methods for assessing soil quality so that remedial actions can be taken in an analytical way. Larson and associates addressed the potential of precision agriculture for environmental protection, but primarily confined their discussion to nutrients, pesticides, and soil erosion.

Water quality is evaluated by the amount of harmful chemicals and sediments the water contains. Harmful chemicals are those that damage biological organisms. For many chemicals, the U.S. Environmental Protection Agency (EPA) has set maximum safe limits. During the past two decades, increasing concern has been raised about groundwater being contaminated with chemicals and surface water being contaminated with both chemicals and sediment as an impact of agricultural practices.

According to the U.S. Geological Survey, nitrate-N levels in groundwater were above the 10 milligrams per liter limit determined safe by EPA in at least 25 percent of sampled wells in 87 counties in the United States, mostly in the Midwest. The U.S. Department of Agriculture (USDA) estimates that nearly half the counties in the United States have groundwater supplies vulnerable to pesticide and nitrate-N contamination, potentially affecting 54 million people who rely on these sources for drinking water.

An estimated 221 million pounds of pesticides were used on cropland or pasture in the Mississippi River Basin in 1991. Five herbicides accounted for about 63 percent of herbicide use in the Mississippi Basin. These herbicides are used mostly on corn, soybeans, and sorghum. Of the herbicides used on corn, atrazine exceeded allowable drinking water contaminant levels in 27 percent of samples from smaller tributaries and the lower Mississippi and Missouri Rivers. Atrazine was detected in all of 147 streams during April-June of 1989-1990. Conservative estimates indicate that less than 1 percent of pesticides applied each year reach the Gulf of Mexico. As in the case for plant nutrients, many improved agricultural management practices are on the shelf waiting to be used, or nearly ready to be used.

Since 1990, the use of herbicides and insecticides has been level, but the use of defoliants and soil fumigants has increased. Pesticide runoff potential for 13 crops is greatest in the Cornbelt, while leaching potential is greatest in the Southeast, Illinois, and Michigan. Potential for phosphorus from animal manure to meet or exceed plant uptake and removal on nonlegume-harvested cropland and hayland is scattered; significant areas exist in the Southeast and Southwest.

Because agriculture is believed to be the largest source of nitrogen and phosphorus in the Mississippi River, agriculture has often been pinpointed as an industry that needs to use corrective measures. The mean annual concentration of nitrate at the mouth of the Mississippi remained nearly constant through the 1950s and then doubled over the last 35 years. There is some indication that the concentrations have levelled off in recent years. Of the total nitrogen flux into the Gulf of Mexico via the Mississippi each year, 31 percent comes from the upper Mississippi, 23 percent from the lower Mississippi, 22 percent from the Ohio, 8 percent from the Central Mississippi, and 6 percent from the White/Arkansas river basins. Silicate concentrations have been reduced 50 percent over the last 35 years. Reliable data for phosphorus over this period are not available.

The data concerning plant nutrient increases in surface water over such a wide area as the Mississippi Basin is cause for concern, and elevated concentrations of nutrients occur in many other drainages across the nation. It seems prudent, therefore, that agriculture re-examine its cultural practices with regard to fertiliser and manure use and aggressively develop educational programs for use of best management practices. Many practices are available and only need implementation.

Soluble phosphorus concentration in runoff water is linearly related to the available phosphorus in eroded sediment. Maintaining the available phosphorus content in different areas of a field at some reasonable level, consistent with economic crop growth, would limit the concentration in surface water. Larson and associates gave examples of fields where the available phosphorus contained in surface soils varied widely. For example, in one field in Minnesota the available phosphorus ranged from 0 to 110 parts per million. Phosphorus content in runoff water can also vary with the method of phosphorus management (placement, time of application), residue management, and tillage.

Application rate and method of management (timing, wind speed, tillage, residue management, application equipment, etc.) can also affect the amount of herbicide in surface runoff and that leaching to groundwater. A promising approach to addressing this problem is weed mapping followed by selective herbicide application in those areas where weed infestations are serious.

Understanding Time and Space Scales

Precision agriculture requires an understanding of time and space scales. Time scales are critical because operations occur when they will benefit the crop most. Space scales become a fundamental principle of field management because inputs and cultural practices are varied with soil type, pest population, or crop maturity. The challenge is to determine how to use time and space scales to advantage in developing an improved understanding of agricultural management. To fully achieve the goals of precision agriculture, management must be applied in a space and time context.

Shelby Fleischer addressed the issue of time and space during the workshop using insect populations within fields as the model. Insect populations vary throughout time because of the different stages of development. Monitoring has become a critical component for quantifying insect populations within a field. Integrated Pest Management (IPM) systems for insects have used the principle of economic threshold to determine and to quantify the potential economic impact on the crop. Monitoring to determine the insect population becomes a critical component of this program. Adoption of IPM principles has led to a reduction in insecticide application. Presumably, there is reduced environmental impact as well.

Differential application of pesticides over time creates potential areas of spray versus no-spray. This, in turn, creates areas of refuge for insects and a change in the spatial variation of insects across a field or among fields. Sampling these areas over time provides a picture of insect population dynamics that will lead to better management decisions. The techniques used to understand the spatial dynamics of these populations include spatial statistics that can be layered against temporal statistics. The application of these methods to quantifying the spatial and temporal patterns across a field may help develop better management methods.

Experiments in space and time

The challenge of monitoring in space and time is important to document the changes that are naturally occurring within a field. To fully realise the potential impact of precision agricultural principles on environmental quality, however, will require the design and implementation of experiments in space and time. Dan Long provided workshop participants an overview of the current statistical methods that can be used to design experiments. A major challenge is to overcome the problems associated with moving from small areas with uniform soils to large fields. Precision agriculture studies will require an approach that uses geographical data statistical analysis. The main purpose of field experiments is to compare effectiveness of different treatments. Precision and accuracy are paramount, but valid assessment of error is also important. Yield is influenced by non-treatment factors, such as pests and soil fertility. If ignored, this extraneous variation leads to erroneous comparisons. Proper field design and statistical analysis will help minimise this problem.

Classical methods for controlling extraneous variation include replication, blocking, and randomisation. Replication increases the number of observations.

Blocking provides local control over soil gradients, thus increasing precision. Randomisation ensures the unbiasedness of treatment comparisons. An assumption that blocking will control soil gradients is unsound if fertility gradients exist within blocks. An assumption that randomisation will neutralise this variability is also untenable.

In precision agriculture research, plots take up large areas when they are established across entire management units, with full-scale planting and harvesting equipment. Estimated plot effects, or plot errors, against their locations within the field will likely show substantial spatial structure or autocorrelation. Classical statistical theory that ignores this autocorrelation can lead to imprecise results and invalid assessment of error.

Classical approaches to controlling non-treatment variance include a variety of methods. Non-treatment variance can be controlled through experimental designs that employ local blocking structures within which gradients are relatively constant. The most efficient designs are impractical because of restrictions on the number of treatments and replicates. Tradition and ease of management force use of the less complicated randomised complete block design. Spatial statistics provide an alternative to blocking designs for use over large areas.

There are a number of techniques that include terms in a model to explain Non-treatment trends in data. Analysis of covariance and least squares smoothing are two methods. Other techniques that indirectly model the correlated structure in error include nearest neighbour analysis and spatial autoregression. There are techniques that directly model the correlated structure in error, and some of these are the mixed models found within SAS, which is appealing because the error term is modelled directly. But success depends upon determining the correct form of the error variance-covariance matrix.

There a number of additional issues facing empirical analysis of data collected in precision agriculture studies. Because precision agriculture studies are often large, field-scale studies, there is a need to determine the most appropriate way to lay out field-scale trials. The spatially balanced, incomplete block designs ensure that treatment comparisons are made over equal distances. Comparisons over equal distances effectively neutralise plot-to-plot correlation, and the restriction on randomisation increases the number of replicates.

Using a series of replicated, small-plot experiments located within management zones of fields assumes an a priori knowledge of how to create management zones; it also requires use of small-plot equipment. This may negate some of the advantages of large-scale field studies. Replicated strip trials with long and narrow plots oriented lengthwise in the field allows testing application rates versus management systems. There remains the concern, however, of what constitutes the plot or experimental unit, the ease of management, and the quality of yield data.

7

Water Management for Precision Farming

Agronomic practices are varied within a field to match locally varying conditions under precision farming. This concept, when applied to irrigation water management, requires that we look at those conditions which could vary locally, and which could influence water management strategy. It should be noted that we are only concerned with variations that occur on a scale smaller than the smallest manageable block.

For example, there may be tremendous variability over 320 acres. But if the fields and irrigation systems are divided so that individual 40-acre blocks can be managed separately, only the variations within the 40-acre blocks may require special treatment.

Plant Conditions

Although we tend to think of commercial agriculture in terms of fields planted to a uniform crop, there are times when certain crop factors may vary within a field.

Growth of Multiple Crops

Multiple crops may be grown within the same field: Although not common in California or the West, occurrences of this phenomenon precipitated some early instances of what might be called precision irrigation in today's terminology. An example is the use of center pivot irrigation systems to irrigate large fields comprised of smaller sub-fields, each with its own crop. The sub-fields may be divided by radial lines, like a pie, with each sector containing a different crop. Or, the sub-fields may be divided by circumferential lines, like the rings on a target. Combinations may also be used. Applications engineers have devised special systems to provide the benefits of large scale center pivot irrigation to collections of smaller plots by arranging these plots in this fashion.

In one project, the sprinklers on the center pivot were divided into an inner and outer set. The inner set ran along the inner 70% of the machine's length, and covered an inner circle of 50% the total irrigated area. The outer set ran along the outer 30% of the machine's length, and also covered 50% of the total irrigated area. The machine could operate with one or the other or both sprinkler sets spraying water. Further subdivisions were made radially, and the speed of the machine varied the watering time. This system provided the water managers with considerable flexibility and control. With judicious placement of crops according to system capabilities and requirements, proper water management could be achieved for multiple crops throughout the field.

More recently, site-specific water applications have been accomplished by fitting up to three, independently valved, 30-foot long manifolds at each location along a center pivot. Each manifold was designed to apply 1, 2 or 4 times the minimum application rate. By operating various combinations of manifolds at each location, and adjusting the center pivot's rotation speed, variable irrigation depths can be applied to blocks through out the irrigated circle.

Changing Plant Variety and Spacing

In a vineyard, for example, a single field may contain blocks with different conditions. The variety, spacing, or trellis/pruning regime may vary from block to block. The irrigation system or operation may need to be adjusted to accommodate the block-to-block variations. If vine spacings vary between blocks, then furrow irrigation frequency or hours of operation may need to vary also to compensate. In the case of drip irrigation, changing drip emitter discharges may provide sufficient compensation. In this case, the field may be irrigated as a whole, and wouldn't require a separate system for each biock.

Varying Plant Size

Particularly for permanent crops like trees and vines, the cumulative effect of past practices may have resulted in trees of varying size throughout the field. Or replacement of damaged may result in newer, and hence smaller, plants at some locations. A recent project categorised citrus trees in a field as either large or small, and supplied water to large and small trees independently by using two independently valved irrigation lines down each tree row. One line supplied water to microsprinklers adjacent to the large trees, while the other one similarly supplied the small trees

A similar, but more complex system has been used to differentially irrigate a mix of indoor, ornamental potted plants. A complicating factor was that the mix of plants was expected to change from time to time. Plants were identified as belonging to one of four categories based on type, size and water requirements. The irrigation system consisted of four independently valved sub-systems, each with supply lines running through our the area to be irrigated. A line from each sub-system was placed along each plant row, and each location was fitted with drip outlets from all four lines. The outlets could be manually opened or closed. Whenever a plant was placed or replaced at a particular location, the drip outlet to the appropriate line was opened. The irrigation time and frequency for each sub-system was varied to match the needs of plants in that category.

Plant size may also vary in a way that is not known a priori. This is particularly true for annual crops, where to identify and map undersized plants prior to the installation of the irrigation system may not be feasible. In this instance, a precision farming would require sensing the size of the plant, using sensors placed or brought to each location, and adjusting the agronomic practice based on the sensor readings. For precision irrigation, smaller water applications would be delivered to the smaller plants.

A location could be an individual plant or contiguous group of plants. If the agronomic practice were delivered by vehicle, then an optical sensor on the vehicle could provide the information needed to adjust the delivery rate. Moving irrigation systems, such as center pivot or linear move systems could use such an approach, with several sensors located along the length of the moving machine. Controllable areas might be on the order of hundreds of square feet: the 30 foot manifolds mentioned above, combined with a spay pattern width of 10 feet yields a control area of at least 300 square feet. So sensors need the ability to survey and integrate their readings over such areas.

But for stationary irrigation systems, one, or even a several, sensors and information relay devices can't cover the entire field. Instead, one sensor and information relay is needed for each location. This is not presently feasible for the large plant populations associated with most annual crops.

Influence of Soil Properties

It is well known that soil properties may vary on a geographic scale much smaller than the commercial agricultural field scale. Two important soil properties that influence irrigation and water management are infiltration rate and available water holding capacity (AWHC). These can be particularly important for surface irrigation methods, wherein the soil plays a key role both in transporting the water through out the field, and in controlling its infiltration into the plant root zone.

Variability of Infiltration Rate

Infiltration rates may vary with location in either a systematic or random fashion. The most common cause of systematic variation is differences in soil compaction. Wheel rows tend to be more compacted than non-wheel rows, and consequently have lower infiltration rates. Random variability in infiltration rate is the result of natural differences in soil properties throughout a field.

The figure below illustrates the variation in infiltration rates at one site in Oklahoma. The soils are generally loam and silt loam soils. The samples for both the wheat and grass data were taken from 4 sub-plot each of about 4 acres in size. The sub plots were adjacent, but separated by about 30 feet. Thus the variability shown for each crop is experienced over an area of less than 20 acres.

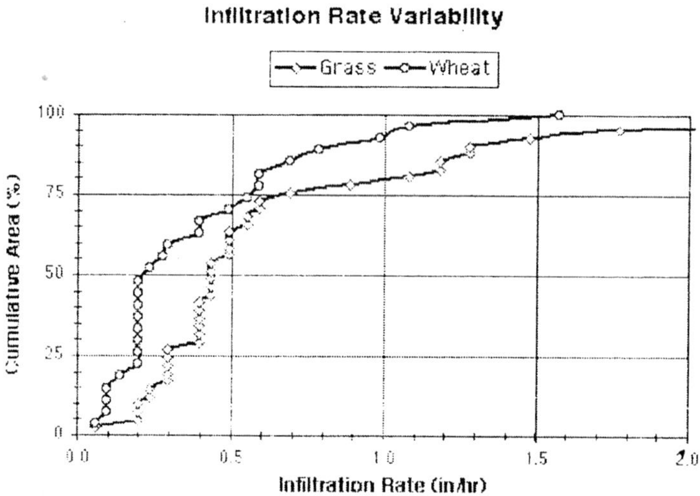

Figure 1.

Variability in soil infiltration rate as measured by double-ring infiltrometers.

The median infiltration rate for the wheat area was about 0.2 in/hr. But about 15% of the area had infiltration rates of half that rate or less, and about a third of the area had infiltration rates of twice that rate or more. The median infiltration rate for the grass area was about 0.4 in/hr. But about 10% of the area had infiltration rates of half that rate or less, and about 25% of the area had infiltration rates of twice that rate or more.

Available Water Holding Capacity (AWHC) Variability

The amount of water that is held by the soil and is available to the plants is dependent on soil type. Typical values for a range of soil types are tabulated here:

Soil Type	*Available Water Holding Capacity*
Fine Sand	0.86
Sandy Loam	1.07
Clay Loam	1.68
Loam	1.78
Clay	1.87
Silt Loam	2.08

The silt loam holds more than two and a half times more water that the fine sand. This is an important consideration when trying to determine irrigation frequency and duration. The following figure illustrates the variation in available water holding capacity at one site in Arizona. The data are from neutron probe measurements made at 32 sites equally spaced through out an area of just over 3 acres. Measurements were made at 8 inch increments and integrated over a 4-foot root zone.

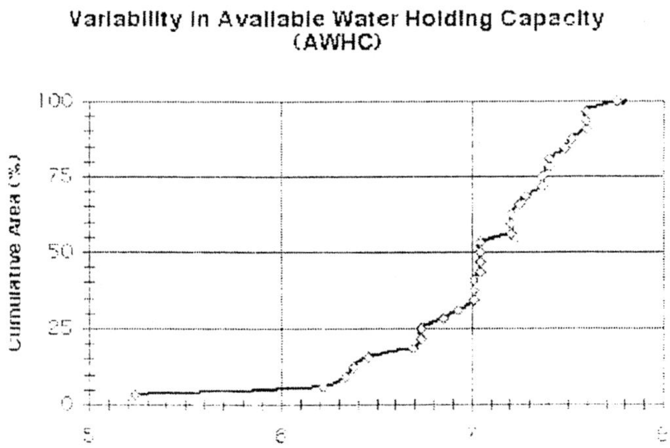

Figure 2

The median AWHC was just over 7 inches for a 4-foot root zone. About 15% of the area had AWHC values of 6.5 inches or less, and about 15% of the area had AWHC values of 7.5 inches or more.

Variable Infiltration Rates Strategies

Possible responses to variable infiltration rates depend on irrigation system type. With surface irrigation methods, the depth of water applied at any one location is controlled by infiltration rate, opportunity time and wetted area. Opportunity time depends on the advance of the water across the field, which in turn is controlled by land grading (slope), surface roughness, and by water inflow rates. For furrow systems, furrow size and shape may affect both advance and wetted area.

Systematic differences in infiltration rate due to wheel row compaction may be minimised by using "torpedoes," or weights drawn through non-wheel row furrows to increase compaction and reduce infiltration rates. An alternate approach is to irrigate the wheel rows and non-wheel rows on different irrigation schedules.

With the possible exception of furrow shape and size, surface irrigation factors affecting infiltrated amount cannot generally be varied on a local scale. And even local variations in furrow size may affect conditions far beyond the local area of the change. For example, narrowing the furrow width to reduce wetted area could compensate for the high infiltration rate in sandy areas. However, narrowing the furrow width would affect water advance, and hence opportunity time at all locations down-slope from the location of the change.

So surface irrigation strategies for minimising the influence of variable infiltration rates is probably limited to reducing the size of managed irrigation blocks, perhaps by reducing the lengths of furrows or border strips. Depending on the length scale over which the infiltration rate changes occur, this may or may not reduce the amount of infiltration rate variability in any one block,. But it will tend to minimise other variations affecting water applications, particularly opportunity time differences.

With sprinkler irrigation, systems are usually designed so that the application rate is less than the infiltration rate. This means that the system's application rate controls water entry into the root zone, not the soil infiltration rate. So soil infiltration rate variability should not require any adjustment in sprinkler system equipment or operation.

High application rate sections of moving systems, such as center pivot and linear move sprinkler systems, may apply water faster than the soil's infiltration rate. But these systems depend on surface storage of temporarily ponded water to prevent runoff until the applied water can be absorbed. Surface storage may be increased by implementing no-till or low-till practices, or by using special tillage equipment to form many small basins to impound ponded water. This latter approach is especially appropriate if the center pivot or linear move systems are fitted with LEPA (Low Energy Precision Application) water emission devices.

With microirrigation systems, the application rate of the emission device controls the amount of water that enters the root zone. While very tight soils can sometimes experience localised runoff problems, even with microirrigation, these are usually controlled by reduced emission rates (accompanied by longer or more frequent irrigations), or localised physical control of the water (small basins).

Strategies for Variable AWHC

Available Water Holding Capacity (AWHC) is the total amount of water in the plant root zone that is available to plants. The strategic response to variable AWHC depends on the irrigation management objectives. For many crops, the irrigation objective may be summarised as "full irrigation." Each irrigation is designed to refill the root zone - the net application amount should equal the amount of water used by the crop since the last irrigation.

The irrigation frequency is chosen to ensure that the soil never gets "too dry" between irrigations. This means that irrigation should be performed before the soil moisture tension reaches a level that would be detrimental to crop growth and yield. As we shall see shortly, the irrigation strategy for locally variable AWHC doesn't necessarily involve locally variable irrigation applications.

For some crops, an irrigation objective other than full irrigation is desirable. These crops need to be stressed at some point in their growth cycle to achieve certain agronomic objectives. An example important to California is the production of wine grapes. Vineyard managers may wish to stress the vines to control vigour and lateral growth of the vines, to maintain berry size, or to affect the chemistry within the fruit (sugar content, etc.).

The vines are stressed by maintaining elevated levels of soil moisture tension. Higher soil moisture tensions can be achieved if the net application amount at each irrigation is less than the amount of water used by the crop since the last irrigation. For these reasons, irrigation regimes associated with desirable stress may be called "deficit irrigation." With these irrigation objectives, locally variable irrigation amounts may be the appropriate strategic response to locally variable AWHC.

Available Water Holding Capacity (AWHC) is the total amount of water in the plant root zone that is available to plants. For a given rate of plant water use (crop evapotranspiration, ETc), field locations with lower AWHC will need more frequent irrigation, and when irrigated will require smaller water amounts when irrigated. But the situation for high AWHC locations isn't exactly the opposite. Field locations with higher AWHC, could tolerate an extended irrigation interval, but don't require one. Field locations with higher AWHC, would require larger water amounts when irrigated only if an extended irrigation interval were used.

In general, the time between irrigations is determined by how much water can be used by the crop before undesirable stress sets in (related to AWHC) and by how fast the crop is using water (ETc). However, the amount of water to add back to the soil at each irrigation is determined by how long it has been since the last irrigation, and how much water the crop has used since then. This means that even in variable AWHC fields, we can manage the irrigations effectively be selecting the irrigation interval appropriate to the those locations with low AWHC, and by setting the irrigation amount appropriate to this interval and the crop water use rate. Those areas with higher AWHC will receive water a little more frequently, and in somewhat smaller amounts than if the irrigations were being managed for them, but no harm is done. They will also remain above the critical stress-generating water level.

Therefore the strategy for locally variable AWHC doesn't necessarily involve locally variable irrigation applications. Knowledge of the variability in AWHC will be used, but only to set the irrigation interval for the field as a whole. When the irrigation is actually done, a uniform amount can be used for the entire field.

A numerical example will illustrate this concept. Consider a crop with a 4 foot root zone using 0.25 inches per day. A full irrigation regime is planned, with the irrigation frequency chosen so that the soil moisture tension never exceeds 125 centibars (125 cb). The crop is grown in a field containing both Sandy Loam and Silt Loam soils. Relevant data for the two soils are as follows:

Soil Type	*Silt Loam*	*Sandy Loam*
Field Capacity (FC)	3.49 in/ft	2.64 in/ft
Permanent Wilting Point (PWP)	1.80 in/ft	1.57 in/ft
Available Water Holding Capacity (AWHC = FC - PWP)	1.69 in/ft	1.07 in/ft
Moisture Content at 125 cb Tension	2.77 in/ft	2.20 in/ft

The amount of water per foot that can be used before the soil reaches the maximum allowed tension is:

FC - Moisture Content at 125 cb

For a 4 foot root zone, the total amount of water that can be used before the soil reaches the maximum allowed tension is:

4 x (FC - Moisture Content at 125 cb)

The maximum irrigation interval will be the this amount of water divided by the rate of crop water use, 0.25 in/day in this example:

[4 x (FC - Moisture Content at 125 cb)]/[0.25 in/day]

The results of these calculations may be summarised as follows:

Soil Type	*Silt Loam*	*Sandy Loam*
Field Capacity (FC)	3.50 in/ft	2.64 in/ft
Moisture Content at 125 cb Tension	2.54 in/ft	2.20 in/ft

Allowable Water Use, Per Foot	0.96 in/ft	0.44 in/ft
Allowable Water Use, Total	3.83 inches	1.75 inches
Rate of Crop Water Use	0.25 in/day	0.25 in/day
Maximum Irrigation Interval	15.3 days	7 days

Irrigating at any interval longer than 7 days will cause the soil moisture tension to exceed the allowable 125 cb in the Sandy Loam portion of the field. But if the entire field is irrigated with an interval of 7 days, the soil moisture tension will remain below 125 cb for both soils.

For both soils, the amount of water used during 7 days will be:

7 days x 0.25 in/day = 1.75 inches

For the 4 foot root zone, this corresponds to:

[1.75 inches]/[4 foot root zone] = 0.44 in/ft

The moisture content just prior to each irrigation will be:

FC - Water Used During 7 days

For the Sandy Loam soil, the moisture content will be

2.64 in/ft - 0.44 in/ft = 2.20 in/ft

which we know corresponds to 125 cb soil moisture tension.

For the Silt Loam soil, the moisture content will be:

3.49 in/ft - 0.44 in/ft = 3.05 in/ft

It turns out that for the Silt Loam soil, this corresponds to only 58 cb soil moisture tension, which is still within allowable limits.

Thus, selecting the irrigation frequency based on the soil with the lowest AWHC, the sandy loam soil in this case, will provide an irrigation regime acceptable for both soils. The change in soil moisture tension between irrigations, assuming a 7-day irrigation cycle, is shown below for both soils.

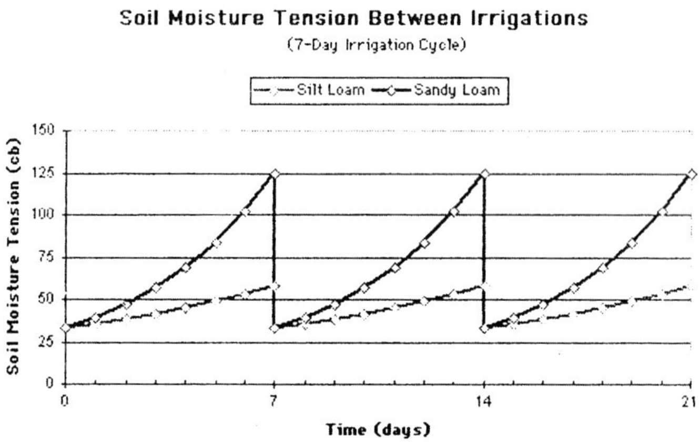

Figure 3

The tension in the Sandy Loam soil increases most rapidly, and reaches 125 cb seven days after an irrigation. The tension in the Silt Loam soil increases more slowly, reaching only 58 cb by the end of the seven day irrigation cycle.

Influence of Rainfall

In areas where significant rainfall occurs during the irrigation season, this strategy may need to be adjusted. The timing and amount of rainfall is uncontrollable. The amount of water stored in the soil from any rainfall event will vary location to location, depending on AWHC. Thus the amount of "effective" precipitation will vary with AWHC. Since irrigation needs only to supply the difference between crop water needs and effective precipitation, this implied locally variable water amounts may be desired. In most areas of California, this shouldn't be a major factor. In the eastern part of the country, though, this may be a significant consideration.

One study looking at the rainfall/AWHC interaction was done in Georgia on soybeans. The field was divided into 5 management zones based on soil properties, with the largest zone being about 12 acres. The study used a soybean growth model to estimate the economics effects of different irrigation regimes. An optimal irrigation strategy was developed for each zone, and computer-tested against 25 years of historical weather data. Irrigating each zone independently, according to its own optimal strategy, produced the highest gross margin. The extra cost to implement independent zone irrigation was not included, however. The strategy of irrigating the field uniformly according to the optimal schedule for the largest management zone produced the second highest gross margins, less than 2% below independent zone irrigation.

When the irrigation objective is to create or maintain a given level of stress throughout a field with varying soil types, it may not be possible to successfully achieve this goal unless different soil zones are irrigated differently.

Irrigation to Desired Stress Level

One consideration involves drying down a field after a period of full irrigation to some desired level of stress (soil moisture tension). Coarse textured soils (with lower AWHCs) hold relatively little water, and will "dry down" to a given stress level relatively quickly. Fine textured soils (with higher AWHCs) hold relatively more water, and will "dry down" to a given stress level relatively slowly. It is not hard to find examples where the coarser textured soils in a field will substantially exceed the desired stress level before the finer textured soils in the same field have reached the desired goal. The problem is even more difficult when different levels of stress are desired at different times during the crop's growth cycle. For instance, one vineyard manager offered the following comments regarding desired stress levels during the year:

Stage of Growth	*Rationale for Stress*
Bud Break to Flowering	Some Stress - To Reduce Vigour
Flowering to Fruit Set	Some Stress - To Reduce Lateral Growth
Fruit Set to Veraision	Increased Stress - To Maintain Berry Size
Veraision to Harvest	Slight Stress - For Flavour Compounds
Harvest to Leaf Fall	No Stress

So to effectively manage a crop where desired stress levels vary throughout the season it may be necessary to use differential irrigation on the different soil zones of the field.

A numerical example will illustrate this concept. Consider a crop with a 4 foot root zone using 0.25 inches per day. After a period of full irrigation, the crop is to be brought to a desired stress level of between 150 and 300 centibars (150-300 cb) moisture tension. The crop is grown in a field containing both Sandy Loam and Silt Loam soils. Relevant data for the two soils are as follows:

Soil Type	*Silt Loam*	*Sandy Loam*
Field Capacity (FC)	3.49 in/ft	2.64 in/ft
Permanent Wilting Point (PWP)	1.80 in/ft	1.57 in/ft
Available Water Holding Capacity (AWHC = FC - PWP)	1.69 in/ft	1.07 in/ft
Moisture Content at 150 cb Tension	2.43 in/ft	2.15 in/ft
Moisture Content at 300 cb Tension	2.06 in/ft	1.95 in/ft

As time passes after the last full irrigation, water extracted for crop evapotranspiration will "dry down" the soil, and increase soil moisture tension. The drying curves for these two soils are illustrated below:

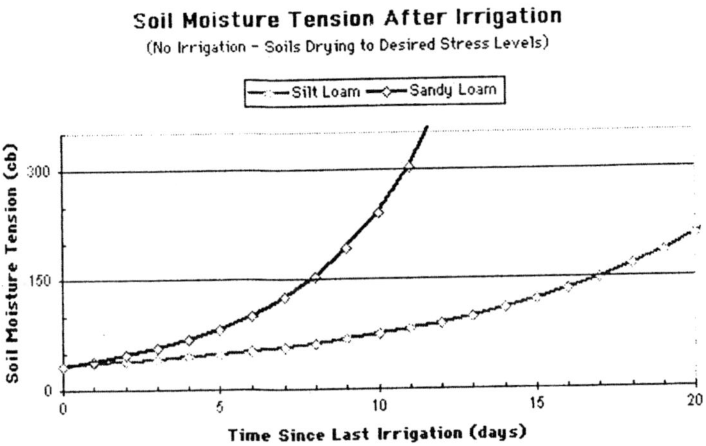

Figure 4

The Sandy Loam soil will reach 50 cb after about 8 days. But at that time, the Silt Loam soil will not be providing sufficient stress for the crop (still only about 60 cb). The Silt Loam soil won't achieve the 150 cb tension level until after about 17 days. But at this time, the Sandy Loam soil will have long since exceeded the maximum tension level of 300 cb (it exceeds 300 cb after about 11 days). Even with the relatively broad band of allowable tensions, irrigation of the coarser textured soil will be required before the finer textured soil has dried down to required tension level.

It is clear that irrigation on the Sandy Loam soil will have to resume no later than 11 days after the dry down period started, but irrigation on the Silt Loam soil will have to wait until 17 or more days have passed. The two soil zones will need to be irrigated differently. You might think that things would be easier if the drying period for the Silt Loam soil was started before the drying of the Sandy Loam soil - but this also would require differential irrigations of the two soil zones.

Maintaining the desired stress level

A second consideration involves selecting an irrigation interval and net application amount that will maintain some desired level of stress (soil moisture tension) once that stress level has been reached. This problem is actually similar to the case of full irrigation. Once a desired stress level has been achieved, irrigating the entire field on the frequency appropriate to the soil with the lowest AWHC will suffice to keep all soils within the desired range of stress.

A numerical example will illustrate this concept. We continue with the circumstances illustrated in previous examples. The a crop has a 4 foot root zone and crop evapotranspiration is 0.25 inches per day. The irrigation regime is to be chosen so that the crop is maintained at stress levels between 150 and 300 centibars (150-300 cb) moisture tension. The crop is grown in a field containing both Sandy Loam and Silt Loam soils. Relevant data for the two soils are as follows:

Soil type	*Silt loam*	*Sandy loam*
Field Capacity (FC)	3.49 in/ft	2.64 in/ft
Permanent Wilting Point (PWP)	1.80 in/ft	1.57 in/ft
Available Water Holding Capacity (AWHC = FC - PWP)	1.69 in/ft	1.07 in/ft
Moisture Content at 150 cb Tension	2.43 in/ft	2.15 in/ft
Moisture Content at 300 cb Tension	2.06 in/ft	1.95 in/ft

The amount of water per foot that can be used while the soil is within the allowed tension range is:

Moisture Content at 150 cb - Moisture Content at 300 cb

For a 4 foot root zone, the total amount of water that can be used before the soil reaches the maximum allowed tension is:

4 x (Moisture Content at 150 cb - Moisture Content at 300 cb)

The maximum irrigation interval will be the this amount of water divided by the rate of crop water use, 0.25 in/day in this example:

[4 x (Moisture Content at 150 cb - Moisture Content at 300 cb)] [0.25 in/day]

The results of these calculations may be summarised as follows:

Soil Type	*Silt loam*	*Sandy loam*
Moisture Content at 150 cb Tension	2.43 in/ft	2.15 in/ft
Moisture Content at 300.cb Tension	2.06 in/ft	1.95 in/ft
Allowable Water Use, Per Foot	0.37 in/ft	0.19 in/ft
Allowable Water Use, Total	1.49 inches	0.77 inches
Rate of Crop Water Use	0.25 in/day	0.25 in/day
Maximum Irrigation Interval	6 days	3.1 days

Assuming the desired soil tension level of 150 cb has been achieved, irrigating at any interval longer than 3 days will cause the soil moisture tension to exceed the maximum allowable 300 cb in the Sandy Loam portion of the field. But if the entire field is irrigated with an interval of 3 days, the soil moisture tension will remain above 150 cb, but below 300 cb for both soils.

Thus, selecting the irrigation frequency based on the soil with the lowest AWHC, the sandy loam soil in this case, will provide an irrigation regime adequate to maintain the desired stress level for both soils.

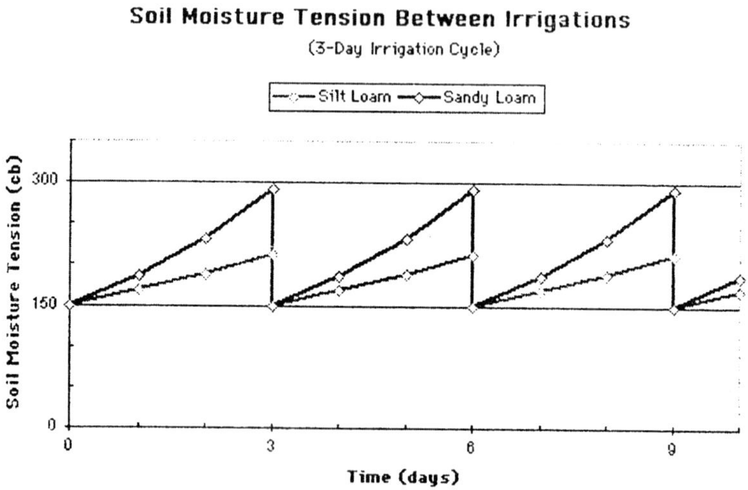

Figure 5

The change in soil moisture tension between irrigations assuming :

(i) The desired tension level of 150 cb has already been reached, and

(ii) a 3-day irrigation cycle is used is shown below for both soils. The tension in the Sandy Loam soil ranges between 150 cb and 300 cb, while the tension in the Silt Loam soil ranges between 150 cb and 209 cb during the three day irrigation cycle.

Basic Tasks for Implementing Precision Irrigation

To implement precision irrigation, two basic tasks must be accomplished:

(1) Zones in the field that are to be irrigated with differing amounts and frequencies must me identified; and

(2) The irrigation system must be designed and installed so that it is capable of sending water to each of these zones on different schedules.

Following are some comments in each of these areas, based on practices employed on some commercial-scale operations in Northern California.

Determining soil types by gridding: A soil sampling grid is established, and the soil is evaluated at a site within each grid cell. At each site, the soil should be evaluated to a depth equal to the rooting depth of the crop. The cost of identifying soil zones by this method can be quite high, depending on the grid interval chosen. The table below tabulates the number of sites to be evaluated in a 80 acre field for various grid intervals.

Grid Size (ft x ft)	*Grid Size (acres)*	*Grid Sites per 80 Acres*
660 x 660	10	8
600 x 600	8.3	10
500 x 500	5.7	14
400 x 400	3.7	22
300 x 300	2.1	39
208 x 208	1	80
100 x 100	0.23	348

The more intervals per acre that a farmer can afford the more accurate the resulting soil profile. However, the cost of creating a soil map by gridding can be quite high for accurate results. So, many farmers choose to use alternate solutions, such as aerial imagery or soil moisture conductivity tests to determine where to do the soil evaluations, thus minimising the amount of soil samples.

Alternate strategies for determining soil sampling sites: One option for determining where to dig soil pits in a field is Aerial Infrared Imagery (IR). IR quantifies the reflectance of the soil surface; therefore IR can distinguish between fine sand, loam and clay soils. Once the farmer knows where the surface soils change, soil evaluation costs can be reduced by selecting only one site per IR-identified soil zone. A word of warning is that just because the surface soils are similar does not mean that the sub soils are the same.

A second method of determining a soil map is by measuring soil conductivity with machines such as a VERIS or and EM-38. These machines measure soil conductivity while being pulled through a field in a grid like pattern. These machines do not identify soil type directly, but they can identify zones within the field where EC values are similar. Presumably, this is related to other soil properties, so EC zones are assumed to indicate boundaries for soil zones. The EC values do not determine the soil hydraulic properties (infiltration rates or AWHC) for the soil zones - these are determined by evaluating soil samples taken from each zone identified.

For fields of existing trees or vines, evaluations of plant growth may be useful. Measurements such as trunk diameter or canopy size may provide indirect indicators of underlying soil properties. Zones of similar values of these indicators may used to identify soil zones, and thus suggest locations for soil sampling.

Vine spacing: In addition to differential irrigation treatment, some vineyards have gone so far as to use different vine spacings with each identified soil zone. The goal is to create a uniform vigour throughout the field by varying vine spacing according to the soil type within each zone. The irrigation system is then designed to supply a different irrigation regime to each zone of soil type and vine spacing.

Irrigation system: Once the zones requiring differential irrigation treatment have been identified, a certified irrigation designer to create an irrigation system design that takes into account not only the topography of the land but the soil and plant spacing characteristics in the identified zones as well. Since the zone boundaries may not match natural subdivisions of field and row dimensions, the resulting design may have to accommodate different soils and vine spacings within the same vine row. This will normally require additional piping and valving, thus increasing irrigation system cost. Before proceeding with such a system, the owner must be convinced that the advantages outweigh the disadvantages.

The manager of one vineyard employing such an approach summarised these consideration as follows:

Disadvantages

- Installation requires a higher degree of management supervision
- Varying vine spacing within a row increased labour costs
  - Labourers were shifted from piece-rate to hourly pay
  - Multiple irrigation sets within one vine row means that labourers must walk up and down rows several times during flushing
- Increased materials costs for the irrigation system (approximately $700/acre in this case)

Advantages

— Reduction in unwanted stress to the vines.

— Increased management flexibility - the system enables not only differential irrigation, but differential fertigation practices as well.

— Reduced water use and pumping costs due to the fact that irrigation is based on smaller and more uniform soil conditions.

— Improved yields from this high value crop should more than off-set the increased installation and labour costs.

Soil moisture sensors and computer automation: Equipment for monitoring the soil moisture status in the various zones may be a desirable addition to the installation. Discussing the zoning of the field and the irrigation block layout with the irrigation designer should help in determining the number and location of soil moisture sensors to be used. Some sensors may be connected to a computer or data logging system to provide real-time soil moisture monitoring. A fully automated system would link the soil moisture sensors through a computer program to the irrigation pump and block valves.

A fully automated system operates under the control of a computer program. Most computer programs allow the farmer to choose what triggers will start and stop irrigation cycles. These triggers could be soil moisture readings, evapotranspiration readings, and/or temperature readings. When one of the start triggers has been reached the computer program sends a signal to the irrigation block that is needing water and opens the block valve and starts the pump. When that irrigation block's stop trigger has been reached the computer program will send a signal to turn off the pump and shut off the block valve.

Most farmers who have installed automated irrigation systems recommend that the system should measure more than just one variable. Additionally, someone must monitor the fields frequently to verify what the computer screen coincides with what is actually what is happening in the field.

8

Point Sampling

Point sampling is one of the basic processing steps in precision farming. In point sampling, issues of spatial interpolation—estimating between sample points such as sampling frequency/pattern and interpolation technique—are the focus of concern.

Let's tackle the first area: the sampling design issues of frequency and pattern. There are four distinct considerations in sampling design:

— stratification,

— sample size,

— sampling grid and

— sampling pattern.

The first three considerations determine the appropriate groupings for sampling (stratification), the sampling intensity for each group (sample size) and an appropriate reference grid resolution (sampling grid). All are closely tied to the spatial variation of the data to be mapped. For example, if your field splits a couple of soil types, you might divide the field into two strata. If one is fairly consistent, you might allocate fewer samples than within the other, more variable soil unit (fig. 1). Also, you might decide to generate a third strata for even more intensive sampling around the soil boundary itself. Or, another stratification possibility is to use your yield map. If you feel the variations in yield are primarily "driven" by soil conditions, then that map could be used to divide the field into high, medium and low productivity areas.

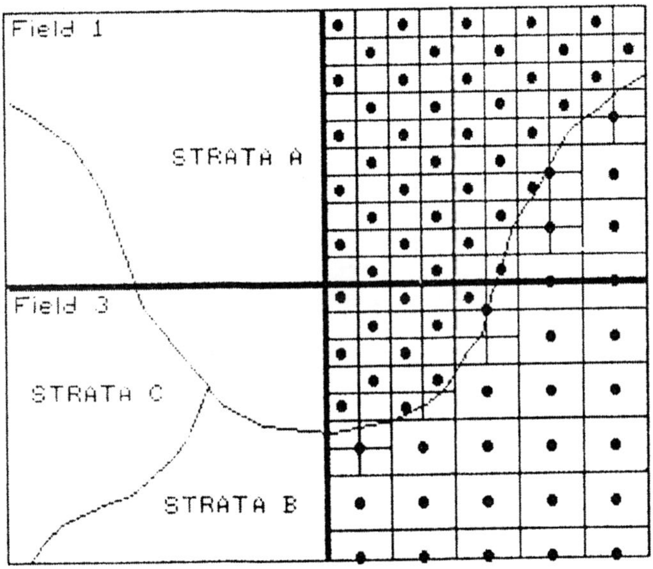

Figure 1. Field stratification

This approach might respond to soil units that are below the resolution of the traditional soil type map or characterised by different nutrient levels. In the past, a single sampling frequency was used throughout an entire state or region. In part, a single frequency was chosen for ease and consistency of field implementation, as well as tradition. With the advent of *GPS* technology, however, variable sample frequencies is a modern alternative—simply navigate to each sample location.

Sampling Patterns

The final step, sampling pattern, identifies individual

sample locations. Traditional statistics tends to emphasise *randomised patterns* because they ensure maximum independence among samples, which is critical to calculating the central tendency of a data set (average for an entire field). However, "the random thing" can actually hinder spatial statistic's ability to map field variability.

Figure 2. identifies five systematic patterns, including a completely random one. Note that the *regular pattern* is uniformly distributed in geographic space. So is the *staggered start* one, except the equally spaced *y*-axis samples alternate the starting position at one-half the sampling grid spacing. The result is a "diamond" pattern rather than a "rectangular" one. This pattern is better suited for generating maps because it provides more intersample distances for spatial interpolation.

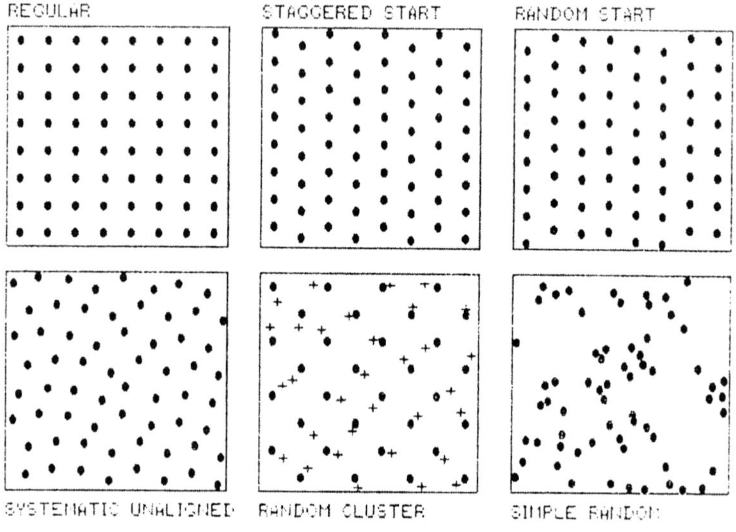

Figure 2. Systematic sampling patterns

The *random-start* pattern begins each column transect at a randomly chosen *y* coordinate within the first grid cell, thereby creating even more intersample distances. The result is nearly a regularly spaced pattern, with "just a hint of randomness." The *systematic unaligned* pattern also results in a fairly regularly spaced pattern, but it exhibits even more randomness as it is not aligned in either the x or y direction. The pattern is formed by first placing a random point in the cell in the lower-left corner to establish a pair of x and y offsets. For rows in the grid, the x offset is held constant while y is randomly varied. For columns, the y offset is held constant while x is varied. The result is random starts in both the x and y axes.

The "dots" in the *random cluster* pattern establish an underlying uniform pattern (every other staggered start sample point). The "crosses" locate a set of related samples that are randomly chosen within the enlarged grid space surrounding each dot. The random subset of points allows a degree of unbiased statistical inference, such as a t-test of significance differences among population means. The *simple random* pattern allows full use of statistical inference (whole field statistics), but the "clumping" of the samples makes it the worst for mapping (site-specific statistics).

From Points to Maps

Spatial interpolation refers to the generation of a continuous map surface from discrete point samples. All spatial interpolation techniques establish a "roving window" that

— moves to a location in a field,

— calculates an estimate (guess) based on the point samples around it,

— assigns the estimate to the center of the window and

— moves to the next location.

The extent of the window (both size and shape) affects the result, regardless of the summary technique. In general, a large window capturing a bunch of values tends to "smooth" the data. A smaller window tends to result in a "rougher" surface with abrupt transitions. Three factors affect the window's extent: its reach, the number of samples, balancing. The *reach*, or search radius, sets a limit on how far the computer will go in collecting data values.

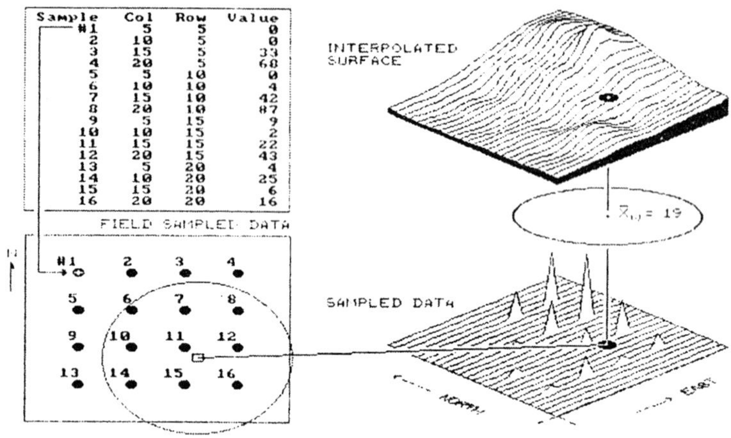

Figure 3. Inverse distance interpolation

The *number of samples* establishes how many data values should be used. If there are more than enough values within a specified reach, the computer uses just the closest ones. If there aren't enough values, it uses all that it can find within the reach. *Balancing* attempts to eliminate directional bias by ensuring that the values are selected in all directions around window's center.

Inverse distance is an easy one to conceptualise (fig. 3). It estimates the value for a location as an average of the data values within its vicinity. The average is weighted in a manner that decreases the influence of the surrounding sample values as the distance increases. In figure 3, the estimate of 19 is the "inverse distance-squared ($1/d^2$) weighted average" of the nine samples in the window. Sample #11 (the closest) influences the average a great deal more than sample #6 (the farthest away). Because this is a static averaging method, the estimated values can never exceed the range of values in the original field data. Also, inverse distance tends to "pull-down peaks and pull-up valleys" in the data. The technique is best suited for data sets with random samples that are relatively independent of their surrounding locations (i.e., minimal regional trend).

The right portion of figure 3 contains three-dimensional (3-D) plots of the point sample data and the inverse distance-squared surface generated. The estimated value in the example can be conceptualised as "sitting on the surface," 19 units above the base (zero). Note that the surface generated by the inverse distance technique is sensitive to sampled locations and tends to put bumps around the sampled areas.).

Surf's Up

The *inverse distance interpolation* technique is fairly straight forward—distance weighted average of nearby sample points. Kriging and minimum curvature techniques give us two more spatial interpolation tools to explore before we can consider which is the best technique to determine residual analysis. Obliquely speaking, the *Kriging* interpolation technique uses "regional variable theory based on an underlying linear *variogram.*" Without immersion into its theory, it is enough to know that there is a lot of math behind the technique. In effect, it develops a "custom" window based on the spacing and directional trends it detects in the sample data. The reach, or size of the window, is determined by intersample correlation.

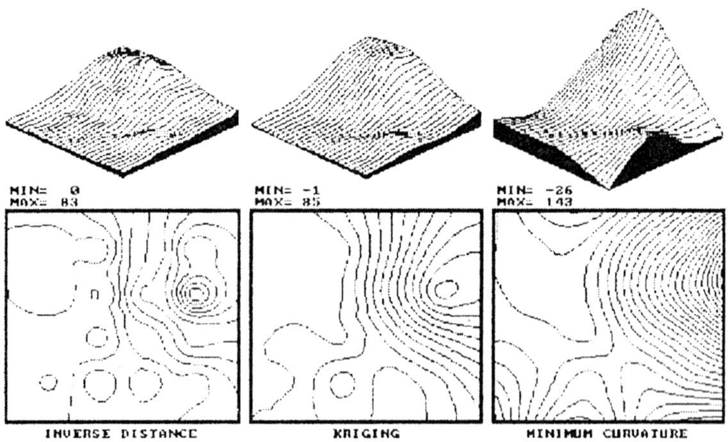

Figure 4. Results of Inverse Distance, Kriging vs. Minimum Curvature interpolation

The summary technique weights the data values, aligning them along the trend's direction more than values opposing the trend. This trend-weighted moving average can result in estimated values that exceed the actual range of the sampled data. Also, there can be unexpected results in large areas without data values. The technique is most appropriate for systematically sampled data exhibiting discernible trends. The center portion of figure 4 depicts the Kriging surface of the sample data described above.

Another technique, *minimum curvature* (mincurve), first calculates a set of initial estimates for all map locations based on localised trends in the sampled data, then repeatedly applies a smoothing equation to the surface. The smoothing continues until successive changes at each map location are less than a specified "maximum absolute deviation," or a maximum number of iterations has been reached. In practice, the process is done on a coarse map grid and repeated for finer and finer grid spacing until the desired grid spacing and smoothness are reached. As with Kriging, the estimated values often exceed the range of the original data values and things can go berserk in areas without sample values.

The right portion of figure 4 contains the plots for the minimum curvature technique. Note that it is the smoothest of the three plots and displays a strong *edge effect* along the border areas that do not contain samples. This technique is appropriate for systematically sampled data and requires samples close to the border. By now you are likely to concede that spatial interpolation is a bit overwhelming and it might seem a bit "fishy." The three plots show radically different renderings for the same set of data. The actual point sampled data had a range from 0 through 87, but none of the interpolated surfaces matches the actual range.

The only thing certain is that there isn't a single best method that covers all situations. But the only good way to determine the "best" map is to empirically verify the results. This involves generating several maps then testing the results against known measurements, aptly termed a *test set*. As in horseshoes and bocci, the closest one wins. *Residual analysis* summarises the differences between map estimates and actual measurements.

Figure 5 displays a test set containing 16 random samples (#17-32) from the data used in the last two sections. The "Actual" column lists the measured values for the test locations identified by the "Col," "Row" coordinates. The difference between these values and those predicted by various interpolation techniques form the residuals shown in parentheses. The "Average" column compares the whole field arithmetic mean of 23 (guess 23 everywhere) for each test location.

For example, the first sample (#17) guesses 23 but was actually 0; therefore, it's off by 23. Note the sign of the residual indicates whether the guesses are too high (positive) or too low (negative). If it were a perfect data world, you would expect the "Sum of the Residuals" to be 0 (just as many high as low misses, in other words, balanced). However, in this case there is a net residual of -17 (indicating a bias to underestimate). The "Average Unsigned Residual" is simply the average magnitude of mistakes, regardless whether more or less than. For the "Average," expect errors of about 22.8 each time you guess. The "Normalised Residual Index" is simply the ratio of the "Average Unsigned Residual" to the "Test Set Average." An index of .81 indicates that the "Average" technique isn't very good.

Now take a look at the residual analysis for the other techniques. All three techniques are considerably better than the whole field average (indices of .18, .12 and .38, compared to a whopping .81), with the Kriging technique best, having a 3.3 "Average Unsigned Residual" and a tendency to underestimate (a "Sum of the Residuals" of -28). The inverse technique is a close second with an index of .18 and a nearly identical underestimate bias (-29). The mincurve technique seems out of the running. However, we haven't considered other potential effects. The

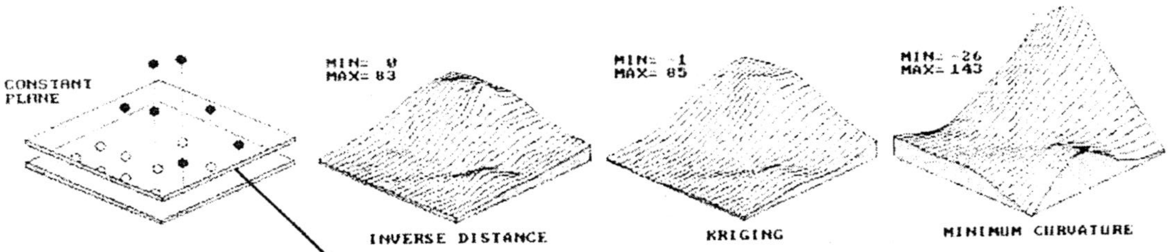

Sample	Col	Row	Actual	Average	Inverse	Kriging	MinCurve
#17*	1	1	0	23 (23)	8 (8)	2 (2)	6 (6)
18*	18	2	48	23 (-25)	42 (-6)	46 (-2)	28 (-20)
19*	23	2	64	23 (-41)	52 (-12)	65 (1)	65 (1)
20*	19	4	65	23 (-42)	54 (-11)	56 (-9)	48 (-17)
21	15	6	34	23 (-11)	33 (-1)	30 (-4)	31 (-3)
22	5	7	0	23 (23)	2 (2)	-1 (-1)	1 (1)
23	9	8	6	23 (17)	7 (1)	1 (-5)	1 (-5)
24	19	11	79	23 (-56)	67 (-12)	70 (-9)	69 (-10)
25	23	13	64	23 (-41)	52 (-12)	68 (4)	90 (26)
26*	4	16	8	23 (15)	8 (0)	7 (-1)	6 (-2)
27	16	17	19	23 (4)	22 (3)	19 (0)	17 (-2)
28*	2	20	6	23 (17)	8 (2)	3 (-3)	-6 (-12)
29	13	22	12	23 (11)	15 (3)	14 (2)	19 (7)
30	22	22	17	23 (6)	19 (2)	20 (3)	7 (-10)
31*	2	24	9	23 (14)	8 (-1)	6 (-3)	-16 (-25)
32	19	24	14	23 (9)	19 (5)	11 (-3)	-7 (-21)
Test Set Average= 28							
Average Estimate=				23	26	26	22
Sum of the Residuals=				(-77)	(-29)	(-28)	(-86)
Average Unsigned Residual=				(22.2)	(5.1)	(3.3)	(10.5)
Normalized Residual Index=				(.80)	(.18)	(.12)	(.38)

Figure 5. Residual analysis table for Average, IDW, Kriging and MinCurve estimates

asterisks identify test measurements outside the set of samples used for interpolation.

To summarise, we have developed a procedure for checking the reliability of an interpolated map, such as a map of soil nitrogen derived from a set of soil samples. It suggests holding back some of the field samples to "empirically verify" the estimated values at the sampled locations. If you don't, then you are simply accepting the map as perfect. The difference between an interpolated "guess" (prediction) and "what is" (measurement) is a *residual.*

A table of all the residuals for a test set is analysed to determine the overall accuracy of an interpolated map. The residual analysis determines that the Kriging interpolation procedure has the closest predictions to the test set measurements. Its predictions, on average, are only 3.3 units off, while the other techniques were considerably more (5.1 and 10.5). The residual table identifies the best interpolated map and just how good it is, but it fails to identify where in the field the map is likely predicting too high or too low. That's the role of an error map.

A Map of Error

Figure 6 shows both a 3-D and a 2-D map of error for the Kriging technique, derived by interpolating its residuals. The posted values on the 3-D surface locate the residual values used to derive the map of error. For example, the value 1 posted for the far corner identifies the residual for test sample #19 at column 23 and row 2; the 4 to its right identifies the residual for test sample #25 at column 23 and row 13 (fig. 5). The dark lines in both plots locate the zero residual contour—right on target.

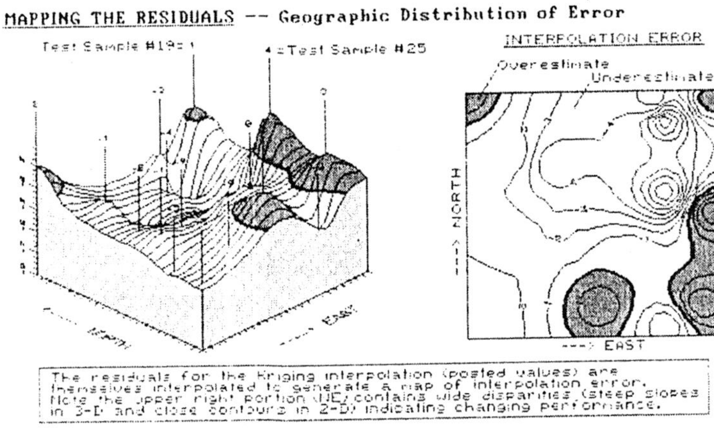

Figure 6. Three- and two-dimensional error maps

It appears that the Kriging overestimates occur along the right (east), bottom (south) and the upper-left (northwest) corner. Underestimates occur toward the center of the field. The upper-right portion (northeast) of the error map is particularly interesting. The two -9 "holes" for test samples #20 and #24 quickly rise to the 1 and 4 "peaks" for samples #19 and #25. These wide disparities within close proximity of each other (steep slopes in 3-D and tight contour lines in 2-D) indicate areas of wildly differing interpolation performance. When developing a prescription for this area, the added information on interpolation error might prove extremely helpful. It certainly takes us beyond simply "pretty maps" toward the mapped data we need in precision farming.

Justifiable Interpolation

The reliability of any map based on point data depends on:

— the existence of spatial dependency within the data,
— the sampling design employed and
— the interpolation algorithm applied.

Spatial dependency within a data set simply means that "what happens at one location depends on what is happening around it" (formally termed *positive spatial autocorrelation*). This simple concept is the foundation for all spatial interpolation.

The *Geary Index* calculates the squared difference between neighbouring sample values, then compares their summary to the overall variance for the entire data set. If the neighbouring variance is a lot less than the overall, then considerable dependency is indicated.

The *Moran Index* is similar; however, it uses the products of neighbouring values instead of the differences. Another approach plots a *variogram* showing the similarity among locations as a function of the distance between sample points. Although the calculations themselves vary and arguments abound about the best approach, all of them are reporting the degree of similarity among neighbouring point samples. If there is a lot, then you can generate maps from the data; if there isn't much, then you are wasting your time. A "pretty map" can be generated regardless of the degree of spatial dependency, but if dependency is minimal, the map is just colourful gibberish; don't bet the farm on it.

Its basic concept involves *spatial autocorrelation*, referring to the degree of similarity among neighbouring points (e.g., soil nutrient samples). If they exhibit a lot similarity, or *spatial dependence*, they ought to derive a good map. If they are spatially independent, then expect a map of pure, dense gibberish. So how can we measure whether "what happens at one location depends on what is happening around it?"

Common sense leads us to believe more similarity exists among the neighbouring soil samples (lines in the left side of fig. 7) than among sample points farther away. The *Geary* and *Moran Indices* consider the closest neighbours. They compute the differences in the values between each sample point and its closest neighbour. If the differences in neighbouring values are a lot smaller than the overall variation among all samples, then a high degree of positive spatial dependency is indicated. If they are about the same or if the neighbours variation is larger (a rare "checkerboard-like" condition), then the assumption of spatial dependence fails. If the dependency test fails, it means the soil nutrient map is just colourful gibberish.

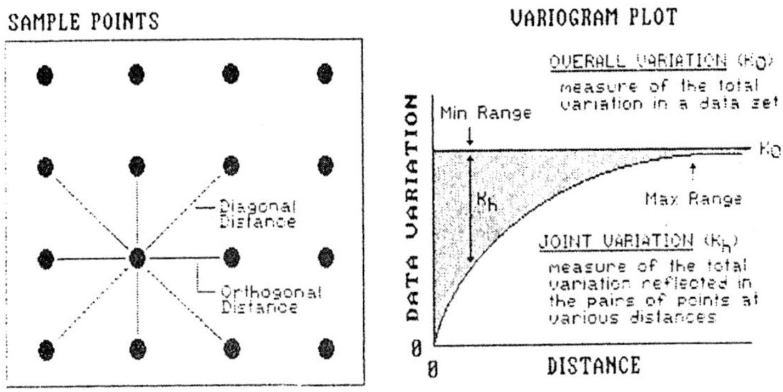

Figure 7. Variogram plot

These indices, however, are limited as they merely assess the closest neighbour, regardless of its distance. That's where a *variogram* comes in. It is a plot (neither devious nor spiteful) of the similarity among values based on the distance between them. Instead of simply testing whether close things are related, it shows how the degree of dependency relates to varying distances between locations. The origin of the plot at *0,0* is a unique case. The distance between samples is zero; therefore, the dissimilarity (data variation = *0*) is exactly the same as itself.

As the distance between points increases, subsets of the data are scrutinised for their dependency. The shaded portion in the plot shows how quickly the spatial dependency among points deteriorates with distance. The maximum range position identifies the distance between points beyond which the data values are considered to be independent of one another. This tells us that using data values beyond this distance for interpolation actually can mess-up the interpolation. The minimum range position identifies the smallest distance actually contained in the data set. If most of the shaded area falls below this distance, it tells you there is insufficient spatial dependency in the data set to warrant interpolation. If you proceed with the interpolation, a nifty colourful map will be generated, but it'll be less than worthless.

Zones and Surfaces

Site-specific farming involves "carving" a field into smaller pieces that better represent the unique conditions and patterns occurring in the field. Two fundamental approaches are used: management zones and map surfaces. *Management zones* use a farmer's knowledge, air photos, terrain features, yield maps or other factors to identify discrete areas considered homogeneous. Sampling, analysis and management decisions are undertaken for each distinct zone—as if they were separate, mini-fields.

Map surfaces, on the other hand, treat a field as a continuous surface by partitioning it into thousands of grid cells that track gradual transitions throughout the field. The resulting grid spaces represent tiny snippets of the field, and information is assigned to each to track the pattern of variation.

Both approaches have their advantages, and disadvantages—management zones are intuitive, require minimal data collection and are less expensive to implement. Map surfaces, on the other hand, are not constrained by artificially abrupt boundaries, therefore they better describe field variability and have greater analysis capabilities. "Like all things GIS," an understanding of the nature of the data and the assumptions underlying the approaches provide insight into their differences.

Consider the maps of P, K and pH shown in figure 8. The "Whole Field" representations are characterised as three horizontal planes "floating" at their average values—the same throughout the entire field. The "Management Zones" approach depicts a "plateau" for each of the three zones determined by their averages—the same throughout each zone. Note that Zone 3 shows lower P and K (42 and 384) but higher pH (6.8) than the whole field averages (48, 419 and 6.5, respectively).

Now consider the "Map Surfaces" that were interpolated from the same soil samples used by the other two approaches. In a sense, the approach maps the variance in the data instead of assigning its average everywhere. The maps characterise the field as a gradient—constantly varying. Note the large phosphorous peak in the NE portion of the field (maximum = 140) and the low values in the SE (minimum = 22). The other surfaces also locate areas that are well above and below the "Whole Field" and "Management Zones" averages.

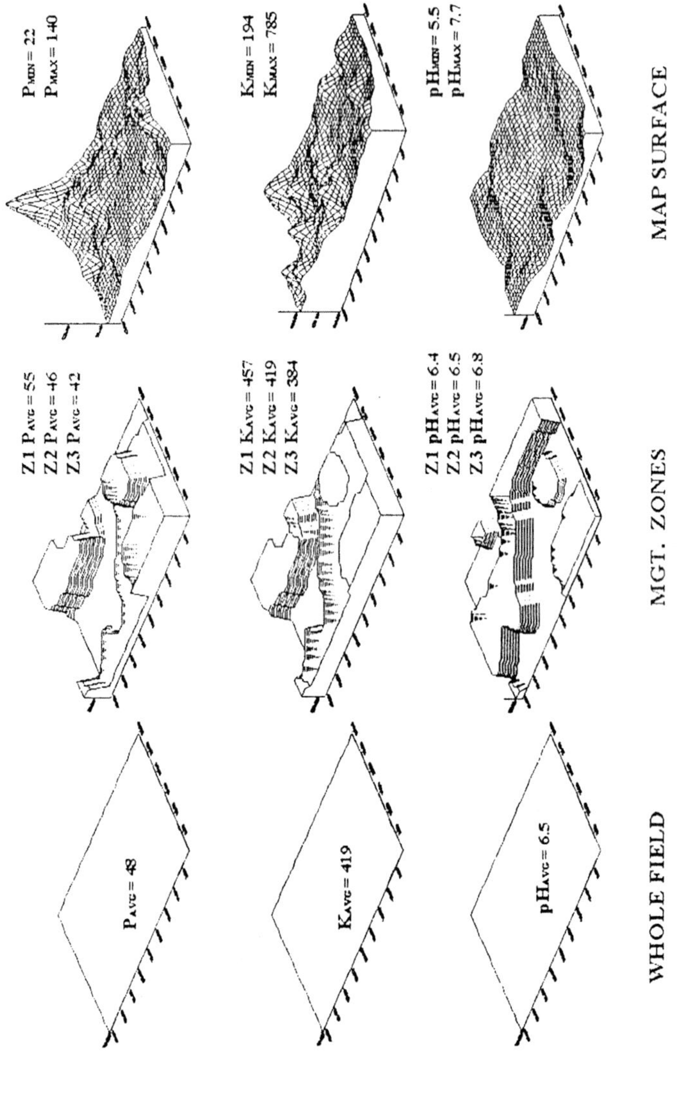

Figure 8. Comparison of soil data maps generated by the different approaches

Table 1. Statistical summary of approaches to subdivide the field

	WHOLE FIELD	MGT. ZONES	MAP SURFACE
Phosphorous (P)	*Avg.= 48 Coffvar= 39%*	*Z1 Avg.= 55 Coffvar= 47%* ← *Z2 Avg.= 46 Coffvar= 24%* *Z3 Avg.= 42 Coffvar= 31%*	*Z1 Min.= 29 Max.= 150* *Z2 Min.= 27 Max = 80* *Z3 Min.= 22 Max= 74*
Potassium (K)	*Avg.= 419 Coffvar= 33%*	*Z1 Avg.= 457 Coffvar= 23%* *Z2 Avg.= 419 Coffvar= 35%* ← *Z3 Avg.= 384 Coffvar= 39%* ←	*Z1 Min.= 295 Max = 625* *Z2 Min.= 261 Max = 727* *Z3 Min.= 194 Max = 785*
Acidity (Ph)	*Avg.= 6.5 Coffvar= 9%*	*Z1 Avg.= 6.4 Coffvar= 11%* ← *Z2 Avg.= 6.5 Coffvar= 8%* *Z3 Avg.= 6.8 Coffvar= 8%*	*Z1 Min.= 5.5 Max = 7.7* *Z2 Min.= 5.7 Max = 7.2* *Z3 Min.= 6.0 Max = 7.7*

The differences among the approaches also show up in statistical summaries (table 2.1). Recall that the *coefficient of variation* (Coffvar) is a frequently used measure that indicates the amount of variation in a set of data—with greater numbers indicating more variation. The "Whole Field" Coffvar's tell us that, throughout the field, there is a fair amount of variation in the P and K values (39% and 33%), but not much for pH (9%). Since the "Management Zones" approach breaks the field into smaller units that are assumed to be more homogenous, it is expected that the Coffvar's for the zones would be less than those of the "Whole Field."

In most cases they are, but the exceptions (identified by arrows in the table) are interesting. They identify zones where the subdividing isn't very good and the averages of the zones are misleading. Note that the data ranges (minimum to maximum value as depicted by the "Map Surfaces") are very large for these zones. For example, zone 1 with a coffvar of 47% has phosphorous values ranging from 29 to 150 (a five-fold difference).

Similarly, the pH range (5.5 to 7.7) for the zone is fairly large. The real problem arises when non-typical conditions align in space, such as the NE corner in zone 1. As both the "Whole Field" and "Management Zones" approaches assume the typical response (average) is everywhere, they miss the combined effects of subtle (and not so subtle) differences from the averages contained in the "Map Surfaces." The result could be significant differences in a prescription for variable rate application of fertiliser. While "Management Zones" is a start toward precision farming and site-specific management, it can fall a fair distance short—it's all in the data and its spatial coincidence.

9

Soil Sampling for Precision Farming

Precision farming offers the potential for improving profits by reducing input costs and improving yields and product quality. By better matching resource inputs to the spatial and temporal variability of environmental factors within farm fields, precision farming may also minimise adverse environmental impacts. The key to efficient and effective site-specific management, is an understanding of the spatial and temporal variability in yield-generating factors and environmental conditions.

In the past, farm fields where managed as single units and managers required estimates of average conditions for soil attributes for the whole field. Precision management requires and understand of variability within the field, and hence sampling is needed to estimate attributes at a finer than whole-field scale. Some significant environmental variables can be measured virtually continuously in space and time or at least on a fine scale (i.e. at fine resolution), and at low cost.

Examples include crop yield (measured with yield monitors), surface spectral reflectance (remote sensing), and electrical conductivity (using mobile sensors).

However, many important soil attributes can currently only be accurately assessed in a laboratory or relate to less accessible subsurface features. Characterisation of these attributes requires sampling. Sampling within-fields provides information on local variability, attribute values, means and ranges for selected locations within the field, provides a basis for variable management (for example variable rate fertiliser application), and facilitates development of management zones.

Soil sampling can be a major expense in precision agriculture. Extra expense can result from taking more samples than are needed for planning within-field management, by taking too few samples to understand variation hence wasting inputs, or by siting the samples inappropriately. Consequently, clear goals and efficient methods are needed to develop a sampling scheme. Soil sampling needs to be planned in time as well as space, because important attributes may vary from year to year, season to season, or even more frequently. Examples of highly dynamic soil properties important to yield and environmental quality include soil nitrate and soil water storage.

Site-specific Sampling

Efficient sampling should aim to address knowledge gaps, rather than exhaustively generate redundant data. Rigid prescriptions for sampling a field for spatially variable management, including recommendations for the number and frequency of required samples and schemes for locating these, ignore local spatial variability and the data and knowledge that the farmers have accumulated specific to their own farm fields. Sampling needs to be tailored to the local variability of a site, and the farmer's information needs.

An efficient sampling design involves analysis of all data layers that can provide information on local spatial variability of the attribute in question. A major objective of information management is the accumulation of knowledge of within-field variability over a number of years and seasons.

The information system developed by the land manager not only informs decisions on inputs, but also provides a rational basis for efficient future information collection. Important data layers include:

- Yield maps
- Soil maps
- Topography maps
- Remote sensing, especially sequential aerial photographs
- The farmer's knowledge of his own field

It is essential to have clear goals for proposed management before gathering large amounts of sample data. The size and nature of management units is also an important consideration.

Types of Site-specific Sampling

Whole field averages

Where a single management regime is applied to a whole field, the aim of sampling is to obtain an estimate of the attribute in question to represent the entire field. Whole-field sampling is frequently based on generation of a single composite sample combining individual samples from a number of points in the field. The number and location of samples is ideally based on knowledge of variability of the attribute (e.g. mean and frequency distribution), the required degree of confidence in the accuracy of the result, and cost. Experience with Ohio soils has lead to a general recommendation that 20 to 30 samples should be taken to make a composite sample within a typical field. For relatively uniform areas, the composite samples should be taken from random locations in the field.

Grid sampling

Grid sampling locates sample points at the nodes or centers of square, rectangular or other regular shaped objects superimposed on the field. Grid locations can be established and maintained using GPS. Grid schemes are convenient to locate and analyse, but, like traditional simple random sampling schemes, may be inefficient when there is prior or accessory data available to inform sampling and application. Typically, a grid density of at least one site per acre is recommended, though frequently fewer samples are taken due to cost. Again, these prescriptions ignore actual local variability.

The rigid shape of the grid means that sample density does not depend on actual variability within the field. As a result, systematic variation (caused by prior linear distribution of fertiliser or manure, for example) may be missed. Modifications to the rigid grid scheme may improve estimates of variability by randomising sample locations within the grid, varying grid size or shape dependent on prior knowledge of variability, or by changing the alignment of the grid throughout the field (systematic unaligned schemes).

Systematic grid sampling permits application of simple techniques to map attributes within fields, to include data in geographic information systems, to analyse relationships between multiple attribute layers, and to interpolate or generalise sample data to finer or coarser grids or map units, as a basis for management decisions. Attribute values needed to generate maps of management units can be developed from the data layers and point samples using geostatistical and other interpolation methods.

Typically, grid samples consist of composites of a number of samples (often 10 to 15 samples), located within a defined distance of the grid node or center (point sampling). Alternatively, grid values can be assessed from composites of random or systematic samples taken to represent each entire grid cell (cell sampling).

Grid sampling may be necessary where there is little prior knowledge of within-field variability.

Geographic unit sampling

In many farm fields, production, environmental impact, environmental conditions and best management practices are geographically related. For example, yield variability may be closely related to soil moisture storage, and soil moisture storage may be closely related to soil series. In other cases, topography may be an important explanatory variable for yield variation. Hence topography or soil maps that partition the field into relatively homogeneous units may be useful as a basis for management or fertiliser application maps. Soil or topographic units, or units derived from yield maps or remote sensing may provide a basis for delineating management zones. Soil sampling may be directed towards developing and understanding these management zones.

By sampling known zones where attributes are expected to be relatively uniform, the number of soil samples required for management decisions may be reduced significantly. Zone sampling facilitates generation of management maps (such as variable rate application maps for fertiliser). Useful map units for directing sampling include large scale soil maps, such as Order1 soil surveys, topographic unit maps, maps developed from sequential yield monitoring, aerial photography and other remote sensing images, and maps produced by "on-the-go" sensors, such as electrical conductivity sensors.

As with whole-field sampling, estimates of attribute values for mapping units may be obtained by compositing a number of samples from within each unit; sample numbers can ideally be determined using knowledge of variability within each unit. While sample numbers and costs involved may be lower for map-unit based sampling, a higher level of analytical skill is needed for efficient application.

"On-the go" sampling

Automatic sensors that collect real-time data on the variability of soil and other environmental attributes are in active development. Ideally, sensors collect data during other in-field operations such as planting, cultivation, pest control or harvesting. Sensors for many soil attributes such as pH, organic matter and nitrogen content are being developed. Sensors for electrical conductivity are currently used widely. Future sensors may mean that many fewer soil samples need be removed form the field.

Smart sampling

As the farm manager's knowledge of within-field variability grows over time, rigid data gathering schemes can be replaced with methods that utilise all available information and the manger's experience to fill knowledge gaps. "Smart" sampling is an iterative approach based on the combination of grid-based, directed sampling and incorporation of prior knowledge to supplement the manager's understanding of within-field variability over time. For example, new samples may be taken in areas of the field where yield variability or in stability over time is high, or in areas where crop response to variable management departs from expectation. Smart sampling may involve the use of models linking the various data sets within a GIS, and knowledge of the site-specific variability.

Soil-Sampling and Testing for Precision Management of Crop Production

Soil-sampling and testing is one of the most important operations in crop production. Traditionally, soil samples are collected throughout the field to get an "average" representation of soil characteristics. Recommendations for fertiliser and lime application rates are based on this collective field representation and used to develop a soil fertilisation and liming schedule. This procedure leads to positive yield gains only if the high and low yielding areas in the field are close to the average. However, experience has shown that a lot of Georgia fields, even small fields, have high and low-yielding areas that are significantly different from the average yield. As a result, a uniform application rate can leave parts of the field with too little or too much fertiliser and lime.

High- and low-yielding areas of the field can be separated into smaller areas and managed independently. Soil samples can then be taken from each smaller area to determine if the soil properties are contributing to the differences in yield and if variable rates of fertiliser and lime may be economically beneficial. Sampling smaller areas will require more soil samples than normally collected for individual fields. If the field does have significant variability in soil properties, the producer may want to consider a variable application rate or addition of soil amendments.

While site-specific soil sampling can improve field management, it must be accompanied by conscientious collection and analysis of data. The new technologies that have elevated precision agriculture into the forefront of farm management cannot offset poor data collection and soils testing techniques. Inaccurate data on soil properties will inevitably lead to improper management decisions.

There are three primary factors that can affect the precision management of soil fertility:

1) collecting proper soil sample cores and the consequences of improper soil sampling,
2) breaking the field into smaller management areas, and
3) differences in test results and recommended application rates between soil-test laboratories.

A step-by-step procedure is then presented on how to use variable yield goals to develop field maps showing variable application rates of fertilisers and lime. Soil samples were collected prior to planting the second year of cotton in the rotation. Samples were collected at two soil sample depths (7.5 cm [3 in] and 15 cm [6 in]) and were sent to two soil test laboratories for analysis of phosphorous, potassium and pH and to obtain recommendations for phosphate, potash and lime application rates. A yield goal of two bales per acre on irrigated cotton was used to calculate soil requirements from both labs.

Developing Soil-Sampling Areas

Before a precision sampling program can begin, a field must be divided into a grid or management zones. A field grid is a uniform pattern of squares, rectangles or other shapes that divide the field. Typical grid areas range from 0.4 to 2 hectares (1 to 5 acres), depending on the size of the field and the number of samples taken. Management zones use plant and soil properties to divide the field into zones that have relatively consistent topography and soil properties. Zone size and shape are dependent on the amount of variability within a field.

Management zones or field grids cannot be created without an accurate field boundary. There are a couple of methods for creating a field boundary. One method is to take a GPS receiver (with differential correction) and laptop computer, with GPS software, to the field and collect field coordinates around the edge of the field. The boundary file along with yield maps and aerial photos are then used to develop management zones or field grids.

These images have the geo-referenced coordinates (latitude and longitude) of the corners and can be used by software packages to create a boundary around a field. Satellite images may also be purchased through Internet providers such as Earthscan Network. The images provided are usually taken every one to two weeks and are available in a variety of formats for creating field boundaries, analysing vegetation growth, determining accurate field acreages, etc. This option is easiest and does not require fieldwork but may create a less accurate boundary. To determine the best option available, contact the software vendors.

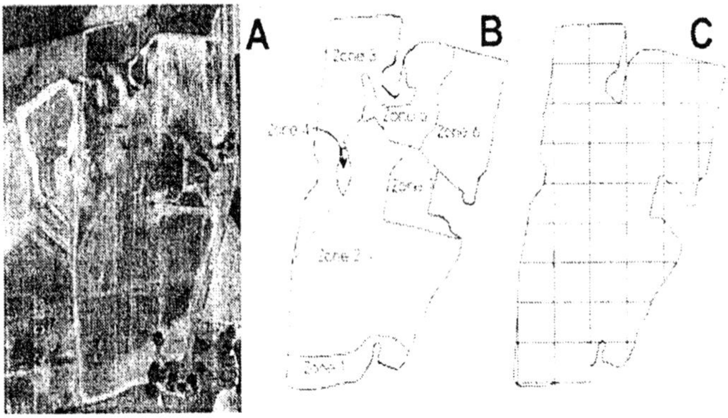

Figure 1. (A) Aerial photograph of 27 ha (67 ac) field six weeks after planting cotton; (B) management zones of field; and (C) 0.8 ha (2 ac) field grids

Several software packages are available to create the management zones or field grids inside the field boundary. An aerial photograph is one of the most useful pieces of information needed to create management zones. An aerial photograph of our 27 ha (67 ac) example field approximately six weeks after the emergence of cotton illustrates some of the history of the field, such as the location of old fence lines and former pastureland, forest or buildings (Figure 1a).

Using the aerial photograph, the previous year's yield map and historical data from the farmer, the field was divided into seven management zones of relatively homogeneous yields, similar history and/or soil types (Figure 1b). In this case, zone 3 was previously pasture and was recently (within 4 years) converted to cropland. Zone 4 is a clay knoll that traditionally has very low yields. Zone 1 is outside the end-gun on the center-pivot irrigation system and zone 7 is a depressed wet-land. These zones take a little effort to draw out and may need "tweeking" the first couple of years but are representative of field variability based on soils data, crop yields and field history.

Grids are usually easier to create and are generally more widely supported by soil sampling and/or fertiliser suppliers and applicators. The number of grids depends on the size of the field and the size of each grid square, rectangle or other shape. However, grids are created using arbitrary boundaries that may have variable soil properties within a grid. Consequently, a variable rate application of fertiliser or lime may still not account for field variability with any improved precision. A software package was used to divide the example field into 0.4-0.8 ha (1-2 ac) rectangles (Figure 1c). Soil samples can be collected at the center of each grid or at the corners where the grid areas meet.

Overall, it is recommended to use management zones to determine fertiliser and lime application rates. It is more difficult to develop the zones, but they usually require less soil sampling than field grids and can more accurately represent the variability in the field. Figure 2 illustrates the example field with variable rates of lime prescribed for each zone based on the soil sample test results and the producer's yield goal. In this case, one yield goal was used for the entire field.

Collecting Soil Sample Cores

Accurate representation of the soil characteristics is extremely important to determining the recommended application rates. In some cases, the soils may be homogeneous over a fairly large depth, while in other cases, soil stratification could largely affect results. A recommendation to sample within ±1 inch of the desired sample depth is recommended here to reduce inaccuracies in desired application rate recommendations.

A GPS receiver and portable computer (laptop or handheld) loaded with one of the software packages listed earlier can provide a method for navigating to soil sample sites. Take samples in management zones from two to three locations within that zone and mixed into one sample.

Lab-to-lab variability can affect soil-test sample results as well as the recommended application rates. When comparing soil samples that were split and sent to two separate laboratories, results between the labs were significantly different for pH, phosphorous and potassium levels in the soil, even though both laboratories used the same soil test procedures (mehlich-1 test for P and K, and pH meter) and analytical equipment.

One laboratory used the University of Georgia crop-response functions and the other used "in-house" developed functions to make application recommendations. Consequently, it is best to remain with one lab for soil analyses and application rate recommendations. This is especially true when the producer is monitoring year-to-year changes in soil characteristics and yield responses. If two laboratories are used, find out from each laboratory what basis was used to make application rate recommendations before making any management decisions. If they do not use the same crop response functions and soil testing procedures, the two lab results and amendment recommendations are not comparable.

Some laboratories will create application maps if they are given a computer file with the field boundary, management zones or grids and the location of the soil sample points. The laboratory will create application rate recommendations in each zone based on the soil test results and the producer's prescribed yield goals. They then create an application map in a computer file for use by the sprayer/spreader truck with a variable rate controller. Consult the soil test laboratory before investing in software packages to ensure compatibility with data formats. Also, if the producer plans on creating his/her own application maps, consult with the company that will be applying the variable rate materials to make sure the software package creates application maps compatible with the variable-rate equipment.

Yield Goals

Consequently, a constant yield goal for a field can be as self-defeating as a constant application of fertiliser and lime. Areas requiring a high fertiliser rate because of poor soil nutrient levels, and a high yield goal may not show a response and cost much more in fertiliser. It is recommended that when looking at a particular management zone, determine a reasonable yield goal for that area and then get fertiliser recommendations based on that yield goal for that zone.

The required tools for soil-sampling and creating application maps includes:

— GPS receiver with differential correction

— Software to store GPS data points (where samples were taken), create management zones or field grids, and create application map for variable rate sprayer/ spreader

— Soil sampling equipment (core sampler, soil sampling bags, etc.)

— Laptop computer or handheld computer to run software and navigate in the field with the GPS

Using these tools, the following is a step-by-step summary of the process for creating a variable rate application map:

1. Develop management zones for taking soil samples and creating variable rate application maps. Utilise aerial photos, yield maps, field history and any other pertinent data known about the field to create -zones within the field. If using an agricultural supplier to variable-rate apply fertiliser and lime, make sure their equipment and software can handle maps created from management zones.
2. Take soil samples in each zone using the soil sampling tools listed above. Always take soil samples at the recommended depth within each zone so soil test results are accurate. Follow the guidelines in the Soil Test Handbook for Georgia appropriate for the particular crop, irrigation, etc.
3. Take samples to a laboratory for analysis. If the field is divided into zones, consider setting different yield goals for each zone that are reasonable, and get recommended application rates tailored to each zone's yield goal. Use the same lab each year for sample analysis and recommendations.
4. Create an application map for the field using the recommendations from the soil test lab in each management zone. Application maps can be created with several software packages and are provided as a service from some soil test labs and agricultural consultants.

10

Remote Sensing and Precision Agriculture

Remote sensing has shown potential for use in agricultural management for a number of years; however, the availability of fine spatial resolution, near real-time data has limited its application in the past. New companies that provide aircraft-based imagery to meet the resolution and temporal requirements for agricultural management are now emerging. The promise of commercially available, high-resolution satellite imagery will also provide additional sources of remotely sensed data.

Advances in precision farming technology (geographic information systems [GIS], global positioning systems, and variable rate equipment) provide the tools needed to apply information from multispectral images to management problems. There is still considerable work to be done before the full benefits of remotely sensed data can be realized, but there are applications that can benefit from this data at the present time.

Remote sensing refers to the process of gathering information about an object, at a distance, without touching the object itself. The most common remote sensing method that comes to most people's minds is the photographic image of an object taken with a camera. Remote sensing has evolved into much more than looking at objects with our eyes. It now includes using instruments, which can measure attributes about objects which unaided human eyes can't see or sense.

Fundamental Properties and Concepts

Electromagnetic Spectrum

All objects including plants and soil emit and or reflect energy in the form of electromagnetic radiation. Electromagnetic radiation travels in waves propagating through space similar to that shown in figure 1.

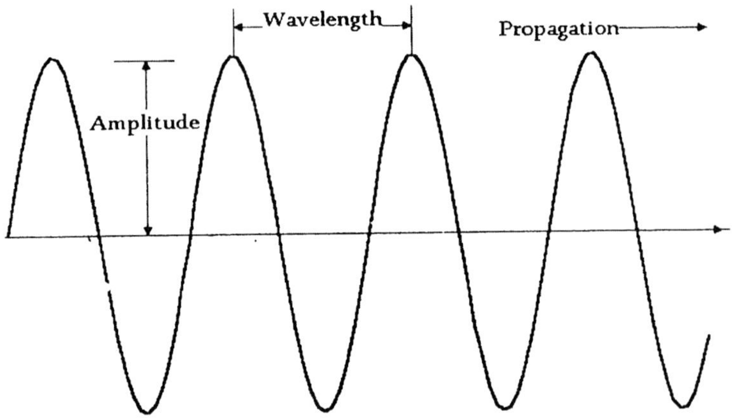

Figure 1. Electromagnetic Radiation

Three major components of these waves are frequency, amplitude and wavelength. Frequency is the number of cycle crests passing a point during a given period of time. One cycle per second is referred to as one hertz. Amplitude is the energy level of each wave measuring the height of each wave peak. Wavelength is the distance from the top of one wave peak to the top of the following wave peak

The most common source of electromagnetic radiation that we are familiar with is the sun. The sun radiates energy covering the entire electromagnetic frequency spectrum as shown in figure 2. Remote sensors act similar to the human eye. They are sensitive to images and patterns of reflected light. A major difference between the human eye and remote sensors is the frequency range of the electromagnetic spectrum that they are sensitive to.

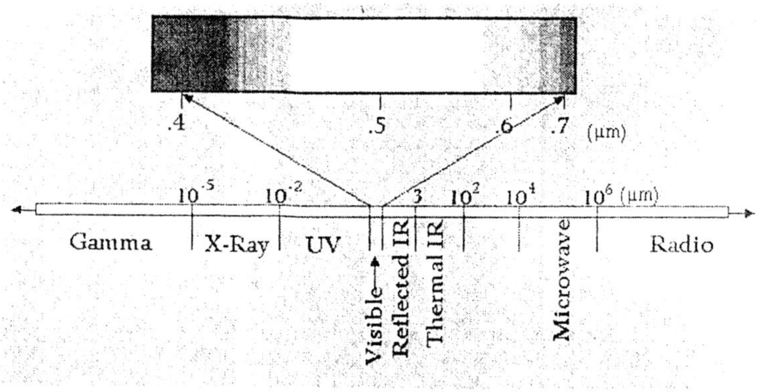

Figure 2. Electromagnetic spectrum

The electromagnetic spectrum range varies from very short wavelengths of less than ten trillionths of a meter known as gamma rays, to radio waves with very long wavelengths of several hundred meters. The electromagnetic spectrum can be sliced up into discrete segments of wavelength ranges called bands, also sometimes referred to as a channel.

It is the sun that most often provides the energy to illuminate objects (figure 3). The sun's radiant energy strikes an object on the ground and some of this energy that is not scattered or absorbed is then reflected back to the remote sensor.

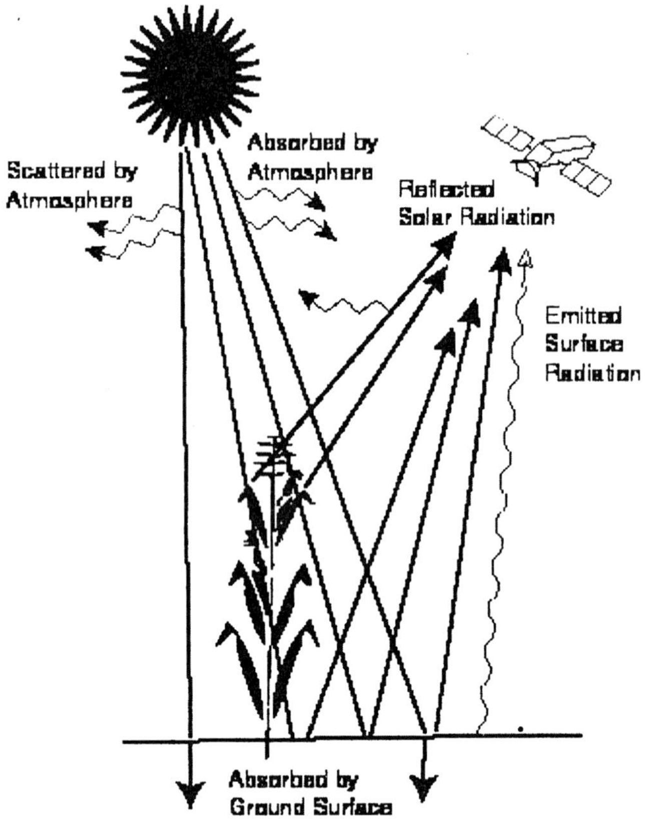

Figure 3.

A portion of the sun's energy is absorbed by objects on the earth's surface and is then emitted back into the atmosphere as thermal energy.

Plants and Electromagnetic Radiation

Leaf structure

The structure of a leaf is shown in Figure 4. The cuticle is a thin waxy layer covering the epidermis cells on the surface of the leaf. Tiny pours in the epidermis layer of cells are called stomata. The stomata are surrounded by guard cells, which cause the stomata to open or close. The guard cells regulate the water evaporation from the leaf and also control the gas exchange between the leaf and the atmosphere.

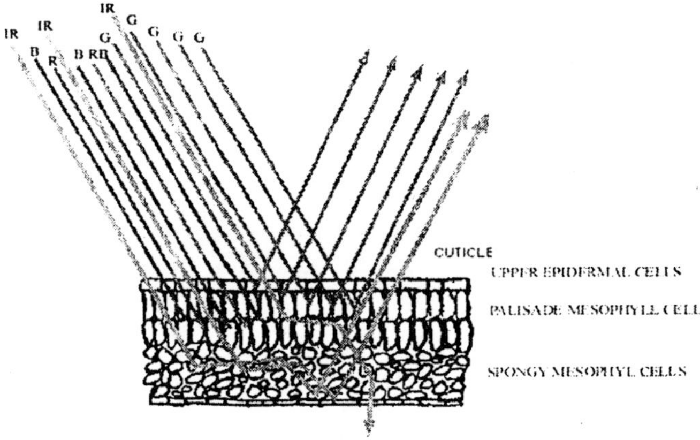

Figure 4. Cross section of a typical plant leaf

The interior layer of the leaf is composed of two regions of mesophyll tissue. This is where most photosynthesis takes place. The palisade mesophyll lies just below the upper epidermis. These cells are elongated, lined up in rows and contain most of the leaf's chloroplasts.

Chloroplasts of most plants contain pigments and two different kinds of chlorophyll. Chlorophyll a is the most abundant and is bluish green in colour. Chlorophyll b is yellowish green in colour and absorbs light and then transfers that energy to chlorophyll a. Pigment molecules within the chloroplasts also absorb light energy and transfer the energy to the chlorophyll. The spongy mesophyll is the leaf lower interior composed of loosely arranged and irregular shaped cells. These cells contain chloroplasts and are surrounded by air spaces.

Spectral response

Chlorophyll primarily absorbs light in the violet to blue and red wavelengths. Green light is not readily absorbed and is reflected thus giving the leaf a green colour appearance. The internal cell wall structure of the mesophyll causes high reflectance of near infrared radiation. Chlorophyll is transparent to near infrared radiation. The sharp increase of reflected energy just beyond the red region of visible light into the near infrared region is referred to as the red edge. Figure 5 shows this sharp reflection increase located around the 0.7 µm wavelength. The location of the red edge is not static throughout the life of a leaf. As the leaf matures, chlorophyll will absorb slightly longer wavelengths in the visible red region. This change moves the red edge shown in figure 5 to the right and is referred to as the red shift.

Environmental stress factors such as drought, disease, weed pressure, insect damage and others stress or injure plants. This stress will cause physiological changes in the plant. Stressed plants will have a spectral reflectance that is different from normal plants at the same growth stage. One example of a physiological change would be the change in the colour of plant leaves due to chlorosis. The yellow colour from chlorosis is caused by the breakdown of chlorophyll. Reflected green will be decreased, and reflected red will increase.

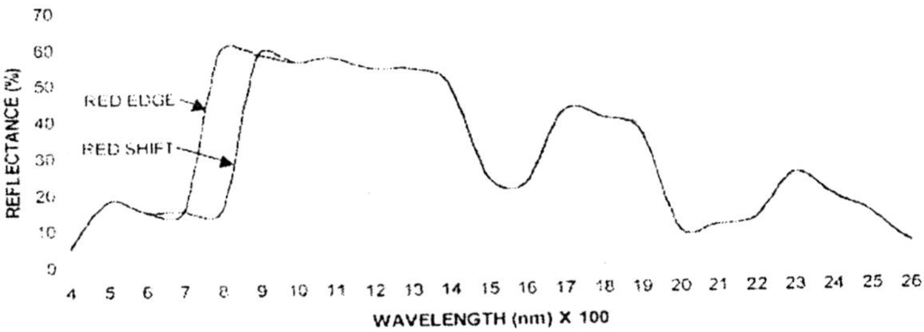

Figure 5. Typical leaf spectral reflectance

The correlation of the different spectral responses observed with remote sensing equipment to the actual condition of the plants, is critical for the accurate interpretation and identification of crop injury and stress.

Sensor Types

Most remote sensors measure and record the magnitude and frequency of reflected radiation from an object. The recorded frequency spectrum data from the object is then compared and matched to the spectrum signatures of known objects thus allowing for the identification and classification of the object on the ground.

Remote sensing from aircraft and satellites use imaging sensors, which measure reflected energy from objects under surveillance. These imaging sensors fall into two general categories, active sensors and passive sensors. Passive sensors monitor only the natural solar reflected light or electromagnetic energy from an object. Passive sensors make up a majority of the sensors in use today. Active image sensors provide their own light or electromagnetic energy, which is transmitted to the object and then reflected back to the sensor. A common example of this type of sensor is radar. Cloud cover in the sky can often block passive sensors from receiving reflected energy from the ground but radar systems can penetrate cloud cover.

Early remote sensing history consisted of photographic images on film taken by cameras. Reflected light received by the camera exposes the film by reacting with the chemical emulsion on the film to create an image in analog format. The images produced are fixed and not subject to very much manipulation unless they are converted into an electronic digital format. Digital images have advantages over analog film images because computers can store, process, enhance, analyse, and render images on a computer screen.

Digital images are images reduced to numbers. The image is made up of numbers, which represent image attributes such as brightness, colour or radiated energy frequency wavelength, and position location for each point or picture element in the image. The smallest size picture elements on a computer screen are called pixels.

Resolution

Remote sensors measure differences and variations of objects. There are four main resolutions, which affect the accuracy and usefulness of remote sensors.

Spatial resolution describes the ability of a sensor to identify the smallest size detail of a pattern on an image. The distance between distinguishable patterns or objects in an image that can be separated from each other is often expressed in meters.

Spectral resolution, is the sensitivity of a sensor to respond to a specific frequency range. The frequency ranges covered often include not only visible light but also non-visible light and electromagnetic radiation. The discrete range of frequency wavelengths that a sensor is able to detect and measure is called a Band. Features on the ground such as water and vegetation can be identified by the different wavelengths reflected. The sensor used must be able to detect these wavelengths in order to see these and other features.

Radiometric resolution is often called contrast. It describes the ability of the sensor to measure the signal strength or brightness of objects. The more sensitive a sensor is to the brightness of an object as compared to its surroundings, the smaller an object that can be detected and identified.

Temporal resolution, is the period of elapsed time between images taken of the same object at the same location. The more frequent a sensor is able to return to an exact specific location the greater the temporal resolution. Several observations over time reveal changes and variations in the object being observed. For satellite systems temporal resolution is described as the revisit period, which refers to the time it takes for a satellite to return to the same area on subsequent orbits.

Image Processing

Once the raw remote sensing digital data has been acquired, it is then processed into usable information. Analog film photographs are chemically processed in a darkroom whereas digital images are processed within a computer. Processing digital data involves changing the data to correct for certain types of distortions. Whenever data is changed to correct for one type of distortion, the possibility of the creating another type of distortion exists. The changes made to remote sensing data involve two major operations: *preprocessing* and *postprocessing*.

Preprocessing

The preprocessing steps of a remotely sensed image generally are performed before the postprocessing enhancement, extraction and analysis of information from the image. Typically, it will be the data provider who will preprocess the image data before delivery of the data to the customer or user. Preprocessing of image data often will include *radiometric correction* and *geometric correction*.

Radiometric corrections are made to the raw digital image data to correct for brightness values, of the object on the ground, that have been distorted because of sensor calibration or sensor malfunction problems. The distortion of images is caused by the scattering of reflected electromagnetic light energy due to a constantly changing atmosphere. This is one source of sensor calibration error.

Geometric corrections are made to correct the inaccuracy between the location coordinates of the picture elements in the image data, and the actual location coordinates on the ground. Several types of geometric corrections include system, precision, and terrain corrections.

System correction uses a geographic reference point for a pixel element such as that provided by the global positioning system. Correction accuracy often varies depending upon the accuracy of the position given by the global positioning system.

Precision correction uses ground control points. Ground control points, which have accurate predetermined longitude and latitude geographic locations, are often used to measure the location error of the picture elements. Several mathematical models are available to estimate the actual position of each picture element based on its distance from the control ground point.

Terrain correction is similar to precision correction except that in addition to longitude and latitude a third dimension of elevation is referenced with the ground control point to correct for terrain induced distortion. This procedure is also referred to as ortho-corrected or orthorectified.

Postprocessing

Digital image postprocessing routines include *image enhancement, image classification,* and *change detection.* These computerized process routines improve the image scene quality and aid in the data interpretation.

Image enhancement techniques include contrast stretching, spatial filtering, and ratioing.

Contrast stretching changes the distribution and range of the digital numbers assigned to each pixel in an image. This is often done to accent image details that may be difficult for the human viewer to observe unaided.

Spatial filtering involves the use of algorithms called filters to either emphasize or de-emphasize brightness using a certain digital number range over an image. High pass filters improve image edge detail. Low pass filters smooth an image and reduce image noise.

Ratios are computed by taking the digital numbers for a frequency band and dividing them by the values of another band. The ratio range can be redistributed to highlight certain image features.

Image classification groups pixels into classes or categories. This image classification process may be unsupervised or supervised.

Unsupervised image classification is a computer-based system that assigns pixels to statistically separable clusters based on the pixel digital number values from several spectral bands. The resulting cluster patterns can be assigned different colours or symbols for viewing to produce a cluster map. The resulting map may not necessarily correspond to ground features that the user is interested in.

Supervised classification is a more comprehensive procedure which uses experienced human image analyst to recognize and group pixels into classes and categories of interest to the user. The analyst picks several samples of homogeneous pixel patterns on the image called training sites. Analysts identify these sites by actually visiting the ground location and making field observations (ground truthing) or by using past experience and skill. The remaining pixels outside the training sites are then matched to the training sites using statistical processing techniques.

Change detection is a process where two images at the same location taken on different dates are compared with each other to measure any changes in physical shape, location, or spectral properties. A third image is then produced which shows only the changes between the first and second image. Change detection lends itself to computer automation analysis. Pixel digital number values are compared pixel by pixel within each frequency band. Computer analysis is most useful when combined with the human analyst's experience and knowledge to interpret the image changes.

Remote Sensing in Precision Agriculture

Precision farming is an integrated agricultural management system incorporating several technologies.

The technological tools often include the *global positioning system, geographical information system, yield monitor, variable rate technology, and remote sensing*. The *global positioning system* ("GPS") is a network of satellites developed for and managed by the U.S. Defense Department. The GPS constellation of 24 satellites orbiting the earth, transmit precise satellite time and location information to ground receivers. The ground receiving units are able to receive this location information from several satellites at a time for use in calculating a triangulation fix thus determining the exact location of the receiver.

Figure 6. VRT Spreader

A *geographical informat.on system* ("GIS") consists of a computer software data base system used to input, store, retrieve, analyse, and display, in map like form, spatially referenced geographical information.

Yield monitors are crop yield measuring devices installed on harvesting equipment. The yield data from the monitor is recorded and stored at regular intervals along with positional data received from the GPS unit.

GIS software takes the yield data and produces yield maps.

Variable rate technology ("VRT") consists of farm field equipment with the ability to precisely control the rate of application of crop inputs and tillage operations.

Remote sensing image data from the soil and crops is processed and then added to the GIS database.

Application of Remote Sensing

Soil and Drainage Maps

Soil maps are also sometimes used to determine management zones. Soil maps are becoming part of the GIS database. The grid sampling technique takes separate soil samples from uniform sized grids laid out over the field. A problem with this type of sampling is the variability that can exist in soil types with in each grid. This variability makes it much tougher to determine soil characteristics within the grid for crop input management purposes. To minimize this problem smaller grids are required which then requires many more soil samples to be take for a larger number of grids. Soil samples can become a major cost of precision farming.

An alternative to grid sampling is targeted or zone sampling. The soil samples are located in homogeneous management zones instead of uniformly spaced grids. The zones are laid out using a process similar to computer based unsupervised image classification. Images obtained from multispectral remote sensors are taken of the vegetated areas of the field. The pixel digital numbers for each band are separated into statistically separable clusters that are classified into homogeneous zones. This cuts down on the soil, terrain, plant growth, and other variability within each area to be managed; thus fewer soil samples are needed for each area.

Except for county soil surveys remote sensing has not gained wide acceptance as a mapping tool for soil characteristics. This is because "the reflectance characteristics of the desired soil properties (e.g., organic matter, texture, iron content) are often confused by variability in soil moisture content, surface roughness, climate factors, solar zenith angle, and view angle".

Subsurface drainage tile lines that have been installed, as long as 50 or more years ago are still partially or totally functional today. Often the existence or location of older tile lines has been lost as landowners die or sell their property. Some states, including Iowa, are now starting to require landowners to prepare and file drainage plats with county recorders when new tile lines are installed. It is desirable to have accurate drain tile maps for maintenance purposes or for the installation of new additional tile line systems. Installation of new tile lines may cut through old tile lines at unknown locations. Building livestock manure lagoons, which cut through old unknown and uncharted tile lines may cause environmental damage from manure leaking through the old tile lines.

Colour infrared ("CIR") aerial photographs have been shown to be an effective tool in locating unknown subsurface tile lines. The image data is digitized for preprocessing and then geo-referenced using ground control points. The CIR photographs show different tones of gray depending on soil type and moisture. By filtering out spectral reflectance differences due to soil type, soil moisture content in dry soils that have a higher reflectance can be identified from lower reflectance wet soils. The resulting image shows were the tile lines are located and whether they are working properly. Normal colour aerial photographs can also be used to locate tile lines. Simple colour photographs offer tile line images similar to CIR but at a lower cost.

Variable Rate Technology (VRT)

One method of controlling variability within the field is VRT. VRT allows the grower to apply the quantity of crop inputs needed at a precise location in the field based on the individual characteristics of that location. Crop inputs that can be varied in their application commonly include tillage, fertilizer, weed control, insect control, plant variety, plant population, and irrigation. Typical VRT system components include a computer controller, GPS receiver, and GIS map database. The computer controller adjusts the equipment application rate of the crop input applied. The computer controller is integrated with the GIS database, which contains the flow rate instructions for the application equipment.

A GPS receiver is linked to the computer. The computer controller uses the location coordinates from the GPS unit to find the equipment location on the map provided by the GIS unit. The computer controller reads the instructions from the GIS system and varies the rate of the crop input being applied as the equipment crosses the field. The computer controller will record the actual rates applied at each location in the field and store the information in the GIS system, thus maintaining precise field maps of materials applied. Although VRT can control inputs applied to crops, it cannot control factors such as soil type, weather climate, and topography that are fixed.

Health Condition of Crops

Remote sensing data and images provide farmers with the ability to monitor the health and condition of crops. Multispectral remote sensing can detect reflected light that is not visible to the naked eye. The chlorophyll in the plant leaf reflects green light while absorbing most of the blue and red lightwaves emitted from the sun. Stressed plants reflect various wavelengths of light that are different from healthy plants. Healthy plants reflect more infrared energy from the spongy mesophyll plant leaf tissue than stressed plants. By being able to detect areas of plant stress before its becomes visible, farmers will have additional time to analyse the problem area and apply a treatment.

Crop Water Stress

The use of remote sensors to directly measure soil moisture has had very limited success. Synthetic Aperture Radar ("SAR") sensors are sensitive to soil moisture and they have been used to directly measure soil moisture. SAR data requires extensive use of processing to remove surface induced noise such as soil surface roughness, vegetation, and topography. A crop evapotranspiration rate decrease is an indicator of crop water stress or other crop problems such as plant disease or insect infestation. Remote sensing images have been combined with a crop water stress index ("CWSI") model to measure field variations. Simple panchromatic aerial photographs have been used to spot irrigation equipment problems. Strips in the vegetation images point to problems with water application rates from defective water nozzles.

Weed Management

One goal of precision farming is to cut crop production inputs, which result in cost and environmental savings. Conventional farming methods apply herbicides to the entire field. Site-specific variable-rate application puts the herbicide where the weeds are.

Aerial remote sensing has not yet proved to be very useful in monitoring and locating dispersed weed populations. Some difficulties encountered are that weeds often will be dispersed throughout a crop that is spectrally similar, and very large-scale high resolution images will be needed for detection and identification.

The use of machine vision technology systems to detect and identify weeds places remote sensors directly on the sprayer equipment. Being close to the crop allows for very high spatial resolutions. Machine vision systems have the ability to be used in the field with the real-time capabilities that are necessary to control sprayer equipment.

Insect Detection

Aerial or satellite remote sensing has not been successfully used to identify and locate insects directly. Indirect detection of insects through the detection of plant stress has generally not been used in annual crops. The economic injury level for treatment is usually exceeded by the time plant stress is detected by remote sensing. Entomologists prefer to do direct in field scouting in order to detect insects in time for chemical treatments to be effective and economical.

Nutrient Stress

Plant nitrogen stress areas can be located in the field using high-resolution colour infrared aerial images.

Management Decision Support Systems

Just having information about variability within the field doesn't solve any problems unless there is some kind of decision support system ("DSS") in order to make VRT recommendations. Russo and Dantinne have suggested the following steps for a DSS:

— Identify environmental and biological states and processes in the field that can be monitored and manipulated for the betterment of crop production.

— Choose sensors and supporting equipment to record data on these states and processes.

— Collect, store and communicate the field-recorded data.

— Process and manipulate the data into useful information and knowledge.

— Present the information and knowledge in a form that can be interpreted to make decisions.

— Choose an action associated with a decision to change the identified state or process in a way that makes it more favourable to profitable crop production.

Cost Benefit Analysis

Remote sensing is just one component of a much larger integrated technology known as precision farming. Precision farming is still an emerging technology with very limited economic success. Precision farming will first have to become an economical management system before the integration and use of remote sensing becomes widely used in agriculture. Early adopters of new farming technology will benefit the most when precision farming does become profitable.

Remote sensing will not be a commonly used technology until precision farming is profitable. Farmers just aren't going to be willing to pay for remote sensing images if its not going to add value to their crop growing operation. Remote sensing can be a very expensive component of precision farming. The cost of a remote sensing images has very little meaning unless it becomes part of a precision farming decision support system.

The following are some examples of the cost of remote sensing images from data providers specializing in agricultural images and decision systems.

Satellite Image Cost

Many satellite companies will not normally deliver remote sensing images directly to the retail customer. Images are sold through authorized retailers. These retailers will take the satellite images and perform post processing steps to make the classified image more useful to the customer.

Aerial Image Cost

Commercial aerial images tend to be higher in cost than satellite images.

Future Developments

Future satellites will have better spatial and spectral resolutions. Launching more satellites will also improve temporal resolution. The delivery time of remote sensing data to the customer will improve. We will someday have real-time satellite remote sensing systems.

Decision support systems will become the main link to convert the spatial data collected into detailed management recommendations at the farmer level. Decision support systems is what will add the most value to remote sensing data for the farmer. The future of remote sensing in precision agriculture will depend upon meeting the needs of the end user, the farmer. Right now remote sensing for agricultural use is still in an early stage of commercial development with unproven economic benefits to the farm producer. The cost of remote sensing data and the other systems associated with precision agriculture will come down to be in line with the benefits received. This is likely to happen in the future as more agricultural information technology companies enter the marketplace.

11

Economics of Precision Agriculture

A common economic decision that farmers make is whether doing something will return more than it costs. This decision is made in a business situation, where returns are measured in dollars. Economics is easy to explain in a business setting because we can use dollar measures. But people make economic decisions happen every day, whether they are in business or not.

For example, deciding how much nitrogen to put on your corn this year, whether to buy a shirt or go out for lunch, or whether to attend a farmer's meeting or spend time fishing are all economic decisions. The concepts may be harder to grasp when costs and returns are not measured in dollars terms, but the concepts are still the same.

Economic decisions involve allocating resources to reach a goal. The goal may be to make as much money as you can, or to grow the largest cantaloupe in the county, or to catch the largest wide-mouth bass caught this period. Whatever the goal, the basic economic decision framework is the same: what are the costs, in dollars, time or effort, compared to the benefits?

Basic Concepts in Economic Decision Making

Economics is a social science that studies how people allocate their limited resources. If resources are not limited, then we do not have to make economic decisions on how to allocate them. For example, air is not limited, so it is not necessary to decide how many breaths to take in the morning and many to save for your afternoon tennis game. Time, on the other hand, is very limited. The more time you spend working, the less time is available for tennis. Money is also limited. The more you spend on food, the less is available for other goods and services.

Utility

People generally allocate their limited resources in such a way that they will receive the greatest return from them. In a business setting the goal may be to maximize profits. However, other goals also may be important. In other words, the greatest return is not necessarily the same thing as the greatest profit. Economists at one time thought that profit was the goal all farmers, but economists now realize that producers and consumers may have multiple goals and objectives. A farmer most likely wants to make a profit, but may at the same time want to be recognized as a community leader, a of champion conservation practices and spend time with the family. This farmer may have multiple goals and objectives, some of which may even be competing with each other while others may be complementary.

Economists refer to the satisfaction received from this mix of activities as "utility". Utility is the satisfaction one receives from consuming a good or a service or engaging in some activity. Profit may determine to a large degree the utility a farmer receives from growing corn, but other factors contributing to the farmer's utility may be having the highest yield per acre, doing a good job and being one's own boss. Many different factors beside economic profit can add to the utility a farmer receives from growing corn.

The idea that farmers may be trying to maximize their total utility rather than their profit is an important concept when it comes to evaluating new technology. Some level of profits is necessary for a business to survive, but a farmer who is doing something that is inconsistent with profit maximization is not necessarily irrational. The farmer may be receiving utility from actions other than profit maximization.

Many studies have shown that farmers are more concerned with minimizing their risk of losing money than with making all the money they can. That is, the loss of utility associated with losing money is greater than the utility gained from a high-risk enterprise. They want to maximize profits from their farming operation, they want to minimize the costs of producing their crops and they want to minimize their down-side risk, i.e., they don't mind making more money than they expected, but will take steps to make sure they don't lose money.

Marginality

An important concept is "marginality". This concept specifies that more resources should be used as long as the marginal benefit from the additional resource exceeds its marginal cost. This concept is easiest to explain with an example. The yield records for corn grown on a sandy loam soil given different amounts of nitrogen fertilizer are reported in Table 1. Without any nitrogen fertilizer the yield was is only 45 bushels/acre. When 20 pounds of fertilizer are applied, the yield increases to 84 bushels. When fertilizer is doubled to 40 pounds, the yield does not double, but it does increase by 26 bushels. As fertilizer is increased by additional 20-pound increments, the corn yield increases, but by smaller and smaller amounts. The maximum amount of fertilizer this crop can use is about 120 pounds per acre. More than 120 pounds per acre and the crop is "burned" by too much nitrogen.

Table 1. Corn Yield and Dollar Values at Different Rates of Fertilizer

Nitrogen Fertilizer (lbs)	*Corn Yield (bu)*	*Value of Fertilizer* ($0.25/lb)	*Value of Corn* ($3/bu)	*Net Return*	*Cost of Added Fertilizer*	*Change in net return*
0	45	$ 0	$135	$135		
20	84	5	252	247	$5	$112
40	110	10	330	320	5	73
60	127	15	381	366	5	46
80	137	20	411	391	5	25
100	140	25	420	395	5	4
120	141	30	423	393	5	-2
140	138	35	414	379	5	-14

Economists call this phenomenon the "Law of Diminishing Returns". Diminishing returns exist in all aspects of agricultural production. An increase in fertilizer causes an increase in yield, but the yield will increase in smaller and smaller amounts as the fertilizer is increased. Give a cow more feed and she will give more milk, up to a point, but the increase in feed will cause ever smaller increases in the amount of milk produced. Diminishing returns holds for all inputs, even information. You can expect an increase in crop yield and/or quality by testing soil moisture and irrigating appropriately. The benefits of testing once a week rather than once a month are likely to be significant. However, there may not be a significant increase in benefits from testing soil moisture every 12 hours rather than every 24 hours.

The economic question is "At what point is the return from the additional yield worth the cost of the additional fertilizer (or feed or information)?" There is no single point, as the "right" amount of fertilizer depends on the prices of fertilizer and corn. Suppose our corn farmer expects to sell corn for $3.00/bushel and nitrogen costs $0.25/pound. At these prices, an additional 39 bushels are worth $117, an additional 20 pounds of nitrogen costs $5.00, and the additional net return is $112. Twenty more pounds of nitrogen cost an additional $5.00 and produce additional 26 bushels of corn, which is worth $78, for an added net return of $73. And so forth. As long as the additional fertilizer costs less than the dollar value of the additional corn yield, it is profitable to apply more fertilizer.

The net return is reported in column 5 of Table 1. The greatest gross margin occurs when 100 pounds of fertilizer are applied per acre. The net return is lower if either more or less nitrogen is applied. This can also be seen by comparing the change in value produced with the additional cost of fertilizer. As long as the additional value produced is greater than the additional cost, the net change in returns will be positive. In real life farmers don't consciously work out tables of added costs and returns, and yet most farmers can tell you doing something differently or adding more of an input would increase yields, but the costs of doing so would out weigh the returns. However, the concept of marginality is central to looking at the economics of a situation. In the example above, we looked at the costs and returns of using more nitrogen on a corn crop "at the margin".

When evaluating precision agriculture systems, you should compare the added costs and the added returns from that system, at the margin.

Relative prices

In the example above, the "right" amount of nitrogen fertilizer was found given the current prices of corn and nitrogen. As prices change, so does the "right" amount of nitrogen. If corn falls to $1.65/bushel, the cost of producing 3 more bushels of corn (from 137 bushels to 140 bushels/acre) is still $5 worth of fertilizer, but the added returns from that fertilizer are only $4.95 of additional corn.

Multiple Enterprises

The concept of marginality still applies even when there is more than one input or output. It becomes a bit more complex, however, when there are two or more outputs, and a limited amount of an input. Suppose you have two possible crops, corn and wheat, with a schedule of added costs and returns from fertilizer as reported in Table 2. The marginality concept tells us to apply five units of fertilizer to both the corn and the wheat. However, if resources are limited to a total of only five units of fertilizer, you cannot set added costs equal to added returns for crops to determine the right amount of fertilizer to use.

The basic marginality concept stills holds, but you will be restricted to applying the limited amount of fertilizer first to the crop with the highest return, then to the next highest return, etc. In Table 2, the first unit of fertilizer should go to the corn. The added return from one unit of fertilizer is greater from corn than from wheat ($117>$80). The second unit, however, should go to wheat. The return of one unit of fertilizer on wheat is greater than the second unit on corn ($80>$78). The third and fourth units should go to corn, and the fifth the wheat.

Table 2. Data Illustrating the Equi-Marginal Principle

Fertilizer (20 lb units)	Cost of added fertilizer	Added Value Produced Due to Added Fertilizer	
		Corn	Wheat
1st	$5	$117	$80
2nd	5	78	50
3rd	5	51	28
4th	5	30	24
5th	5	9	8
6th	5	3	4
7th	5	-9	1

Always allocating a limited resource to the activity with the highest added return is called the "equi-marginal principle". Following this principle ensures the most profitable use of that resource. It can also help to find the most profitable combination of enterprises.

Time is a farmer's most limited resource. Land can be rented, money can be borrowed and labour can be hired, but there are only 24 hours in a day. Good managers understand that the time they spend on a decision is subject to diminishing marginal returns. That is, the added probability of making a good decision increases with the amount of time spent gathering information and analyzing the decision, but there comes a point where more time spent on a decision won't increase the certainty of making the right decision. Good managers allocate their time so that the time spent and the expected payoffs from each decision are about the same.

Time Value of Money

Many inputs, such as fertilizer and chemicals, are used as they are purchased, so evaluating their costs and returns is straightforward. However, evaluating returns from equipment that can last for several years is complicated because the costs and returns don't occur in the same year. The problem is in comparing the value of a dollar today with a dollar in the future. This involves the time value of money.

The central point to the time value of money is that a dollar today is worth more than a dollar tomorrow. The reason for this difference in value is that people normally would rather consume a good today than wait to consume it in the future, everything else being equal. Because of this preference to consume good now, people have to be paid to defer consumption to the future. If you put a dollar in savings account for a year, the bank will return your dollar plus interest at the end of that year. Interest earned on savings, in effect, is payment for deferred consumption.

Interest gives two aspects to the time value of money – compounding and discounting. Compounding means earning interest on an investment and reinvesting the interest earned back into the investment. One hundred dollars invested at 8% would be worth $125.97 at the end of three years (i.e., $100 x 1.08 x 1.08 x 1.08). Discounting is the flip side of compounding. The present value of $125.97 to be paid to you in three years discounted at 8% is $100 (i.e., $125.97/[1.08 x 1.08 x 1.08]). When we compare the costs and returns to investments over time, we often discount the costs and returns to the present and add them together to measure the net present value.

To illustrate using net present value, consider the following example. A farmer buys a calf to background on pasture for a year and then feed for a year to sell as a finished animal. Assume that the calf costs $400 and requires an additional $100 in feed the first year, or $500 in start-up costs, and $300 worth the feed the second year. The finished animal can be sold for $950 at the end of the second year.

If the farmer borrows from the bank to finance buying and feeding the calf, the interest rate charged by the bank should be considered the discount rate. If the farmer uses his or her own money to buy and feed the calf, the discount rate should be the opportunity cost of the money, or what he or she could earn from leaving the money in some other investment.

Table 3 reports the present value of net returns for the above example. If there is no time value to the money (i.e., a "0" discount rate), then the project will yield a $150 net return. At a discount rate (or cost of money) of 8%, the project would yield a net return of $36. And with a discount rate of 11%, the project just breaks even.

Table 3. Present Value of Returns under Different Discount Rates

Present Value of Net Returns

Year	*Costs*	*Returns*	*Not Discounted*	*Discounted at 8%*	*Discounted at 11%*
Start up	500		-500	-500	-500
Start of year 2	300		-300	-278	-270
End of year 2		950	+950	+814	+771
			150	36	1

A business decision that has costs and returns occurring over time involves more than just adding up the costs and returns in one year. Both the costs and returns need to be discounted to take into account the time value of money. The discount rate should also be considered. In general, the higher the discount rate, the lower the net present value of a future return.

Costs of Production

Opportunity Costs

Not all resources that a farmer uses are purchased inputs, with costs determined in the market place. Some inputs, such as own land, fully paid for equipment and the farmer's own labour and management expertise, do not have a cash cost. Nevertheless, these inputs do have a cost associated with their use and they should not be treated as free resources. In each case, these resources have value in their current use as well as in alternative uses. The highest value in alternative uses is called the opportunity cost of the resource.

For example, the opportunity cost of the farmer's time is what the farmer could receive in wages or salary by working for someone else. The opportunity cost of land is the income that could have been received if the land had been put to a different use. Opportunity costs are not direct, out-of-pocket costs, but they are still real costs in terms of what would be lost if the resources were not used effectively. Profits are maximized when the returns to the resource exceeds its opportunity costs; in other words, when the resource is earning more in its current use than in any other opportunity use.

At times farmers tend to treat their own resources as if they were free. They figure if they already own the resource that there is no cost in using or not using the resource. Their own labcur and management expertise is especially overlooked. But all resources have real costs, even if it is only an opportunity cost. Overlooking the opportunity cost of a resource allows for the misallocation of that resource. The ultimate opportunity cost is what the farmer could receive by selling out and investing the money elsewhere.

Every time a farmer drives a tractor across a field, two types of costs are incurred. Understanding the difference between these costs will help you to evaluate the benefits and costs of precision agriculture for your operation.

Operating Costs

Operating costs are the cash costs or out-of-pocket expenses of running the tractor. These costs are primarily fuel and lubrication, repairs and maintenance and labour costs.

Ownership Costs

Ownership costs are incurred whether the tractor is running or sitting in a shed. There are five of these costs:

- — Depreciation
- — Interest
- — Taxes
- — Insurance
- — Housing.

Depreciation is that portion of the tractor you use up in one year. Suppose you buy a $100,000 tractor that has an expected effective life of 10 years. If you use the tractor for 10 years, the cost of the tractor is spread out over those 10 years. In effect, you use $1/10^{th}$ of the tractor each year, or the tractor can be considered to depreciate by $10,000/year.

Interest is the cost to borrow the money to buy the tractor, or the opportunity cost of your money if you use your own savings.

Taxes, insurance and housing are usually a small percentage (2-5%) of the purchase price, but they do add to the cost of owning the tractor.

Insurance and housing combined usually amount to no more than about 2-3% of the value of equipment. However, they should be added as a real cost when one considers a new piece of equipment or machinery.

Precision agriculture systems often involve new and specialized equipment. When comparing the costs and returns from PA, it is important to not confuse the purchase price of the equipment with the annualized ownership cost of the PA equipment. A $10,000 system with a three-year expected lifetime may have ownership costs of $4,000/year. You would be over estimating the system's costs, and subsequently under estimating the system's annual net returns, if you used the $10,000 purchase price rather than the $4,000 annual ownership cost.

Risk and Uncertainty

Farming is a risky business. There are risks that the crop will become diseased, prices will fall, and foreign competition will push you out of business. An additional risk is that the new technology you just purchased in installed will not perform up to expectations, or that the company you bought it from will go bankrupt, leaving you without technical support. The following example of risk and uncertainty is presented to provide a framework for evaluating the risks associated precision agriculture.

Every time you roll an unweighted, six-sided die, you face a 1/6 probability of rolling a 1, or a 2, or a ... 6. Suppose you receive a payoff, say $1, if you roll a 6. Mathematically, the expect value of the game is:

$$\$1 * Prob[1/6] = \$0.167.$$

In words: if you roll the die a lot of times, you would expect to receive the $1 payoff once out every six rolls. On average, for each roll you would receive $0.167. Or, your expected payoff is $0.167.

Now suppose it cost you $0.25 to play the described game. You can use economics to frame your decision; i.e., is the return greater than the cost?

$$\$1*Prob[1/6] = \$0.167 < \$0.25$$

In this case the cost ($0.25) is greater than expected return ($0.167). If your goal is to maximize your return, you would not play the game.

What if the game cost only $0.10? In that case, the return would be greater than the cost, and you would likely play the game.

{What about people who gamble? Why are casinos spreading across the country even though most people know the odds of winning are against them? Think back to utility. Some people enjoy the thrill of gambling – the unknown of winning or losing. People gamble to maximizing their utility – the satisfaction from the act of gambling – rather than from any real hope of making a positive return.}

Farmers play risky games every time they plant a crop. There is a probability that the yield will be an expected amount. Often times there is a probability associated with the price as well. The cost of the crop is known (or should be if the farmer keeps good records). Hence, the farmer can expect profit from a crop to be:

$$Profit = Price * Prob[yield] - cost.$$

Farmers with several years experience have a pretty good idea of the yield they can expect for a given crop on a specific field. Of course, some years' yields will be less than what they expected and other years above expectations. That is the nature of a risky enterprise such as farming. New farmers can get an idea of the probability of having a below average, average, or above average yield from government or private crop advisors, who rely on information from a variety of source to predict how a crop will do in a specific site.

Suppose you were planting a new crop and had little or no idea of what to expect? Economists would say that you are facing an "uncertain" situation. Profits are indeterminate when

$$\text{Profit} = \text{Price} * \text{Prob[unknown]} - \text{cost}.$$

Most farmers prefer a risky situation to an uncertain situation. There are ways to deal with risk, but one faces the unknown with uncertainty.

Value of Information

Precision agriculture has been defined as information technology applied to agriculture. Information has been input in production for years. For example, farmers would know which fields were most productive, what would grow best in which field. Dairy farmers knew which cows would produce more milk if fed more, and which cows would just put on more weight. This intimate knowledge of the production process works well when fields are relatively small and a commercial herd has a score of cows. However, when production decision are made for fields that encompass a section and herds have thousands of cows, intimate knowledge has to be replaced by systematic information in order to have effective decisions made. Information as an input has unique characteristics.

First of all, data is not information. Data has to be screen and edited into a form which provides information. For example, 22 is data but provides no information. Knowing the temperature in Toronto, Canada, is 22°C provides a context for the data, but still provides little information unless you know that 22°C is about 72°F. Information has value when it affects actions or prior beliefs. If you are thinking of visiting Toronto, then knowing that the expected temperature will be 22°C when you are thinking of visiting is valuable information. Further, having a reference point for 22°C, i.e., most Americans know that 72°F is comfortable, may affect your decision to visit Toronto, or re-affirm your prior belief that Toronto would be a nice place to visit in the summer.

The value of information increases with timeliness, accuracy, and lack of bias. If you have planned a business trip to Toronto in two days, knowing what the temperature will be in two days has more value to you than knowing what the temperature was two weeks ago. A forecast of 22°C is more accurate than a forecast of "in the 20's". Lastly, you may expect more positive bias from the Toronto Chamber of Commerce than from Weather Canada.

Yield Gap

All crops have a maximum potential yield, based on the genetic potential of the plant. Livestock have a similar potential genetic yield; e.g., the maximum yield of milk, eggs or meat from a given animal. This maximum is rarely achieved. The reason why can be as obvious as a disease or an insect infestation, or as hard to determine as the lack of water or nutrients at a critical stage of development. The difference between the maximum possible yield and the actual yield can be called the "yield gap". If the plant has received the optimum amount of water and nutrients when required and was not damaged by any pests or other infestations, then yield will be at the maximum genetic potential Y_{max}. However, agriculture is rarely perfect. Most likely, the plant will be damaged by

— less than the optimum amount of water

— less than the optimum soil nutrients

— an insect or weed infestation

— a fungus or other pest infestation.

The yield from a damaged plant, Y_{damage}, is less than Y_{max}. The difference between the two can be called the yield gap. The greater the yield gap, the more the damage costs the farmer in terms lower yield and hence lower revenue. Suppose the crop becomes infested with a blight at time "s". If the farmer does not learn about the blight, he could lose the entire crop. However, suppose the farmer learns about the blight and is able to control the damage. Damage has occurred, but if further damage can be controlled, then the yield will be $Y_{damaged}$, which is less than Y_{max} but better than no crop at all.

The earlier that a farmer knows about damage happening and is able to control the damage, the less the yield gap will be. Mathematically this can expressed as

$$Y = Y_{max} - \int_{5}^{t} D(t)C_t$$

where Y is actual yield, Y_{max} is the maximum potential yield, D(t) is damage occurring over the time period "s" to "t", and C_t is action to control the damage at time "t". If action to control the damage is taken quickly, i.e., at the limit s = t, then damage is minimized. However, the longer it takes for the farmer to take action to control the damage, the greater the damage and the larger the yield gap.

Partial Budgeting

The simple economics of a new technology is that if the returns from using the new technology are greater than the costs of the new technology, then use it. Otherwise, don't. The difficult part is that new technology is often very complex to implement, and determining costs and returns for a new system is rarely simple.

Partial budgeting is a method for comparing the costs and returns from a proposed change in a farm business. It is especially useful for evaluating a specific, limited change with what is currently being done. For example, a partial budget would be a good way to evaluate the costs and returns of a new combine. Suppose the new combine costs less to operate than the current combine, and is expected to harvest three more bushels of corn/acre due to less loss. The partial budget provides a consistent framework for comparing the lower operating costs and increased revenues from the new combine to the cost of buying the combine.

Both a strength and a weakness of partial budgeting is that it is limited to two alternatives. This means that partial budgeting is not a good way to determine which of three crops is the best one to grow, but a partial budget is an excellent way to evaluate whether a new technique or piece of equipment will benefit your operation. The Problem refers to the decision being evaluated, such as buy a new combine or keep the current one.

Additional costs are those costs that will be incurred with the new technique, method or enterprise. Recall that new equipment usually has two types of costs: operating costs and ownership costs. Both are important factors in accurately determining the profitability of an alternative.

Reduced Revues are current revenues that will be lost or reduced should the new alternative be adopted. Not all alternatives will have reduced revenues.

Additional Revenues are those that will be received only if the new alternative is adopted. As with Reduced Revenues, not all alternatives will have Additional Revenues.

Reduced costs are those now being incurred that would longer be incurred if the new alternative is adopted. As with additional costs, both operating and ownership costs need to be considered. For example, replacing an old combine with a new combine means that both the operating and ownership costs incurred from the old combine will be eliminated.

Costs: the additional costs and reduced revenues are the costs of the new alternative. These can be considered the detriments of the new alternative.

Benefits: the additional revenue and the reduced Costs are the benefits of the new alternative.

Net Benefits: if the benefits are greater than the costs, the new alternative has positive net benefits. Any alternative with negative net Benefits should not be considered, as it will cost more than it will return.

Benefit/cost ratio: looks at the relative values of the benefits and costs. An example can best explain why this ratio is important. Suppose you are evaluating two alternatives. Alternative A has $100,000 of Total Benefits and $99,000 of Total Costs. Alternative B has $10,000 of Total Benefits and $9,000 of Total Costs. Both alternatives have Net Benefits of $1000, which at first glance may look profitable. However, the Benefit/Cost Ratio of Alternative

A is 1.01, which means that the alternative is expected to return $0.01 for every $1.00 spent on the alternative, or a 1% return on expenditures, while Alternative B has a Benefit/Cost Ratio of 1.10. Even though the Net Benefits are the same, the B/C Ratio shows that the return from Alternative B is much better. Both Net Benefits and the Benefit/Cost Ratio should be used to evaluate the results from a partial budget.

While some studies have reported positive returns to variable-rate technology (VRT), others have reported costs higher than returns or no significant difference in returns. The economics of PA depend on 1) the system being evaluated and 2) the farm or operation for which the PA system is being evaluated. In other words, some PA system will be economicai on some farms, but by no means will all PA systems be economical on all farmers.

The following are some of the factors that appear to affect the overall feasibility of PA.

— PA is a system, not a single piece of equipment or technology. A GPS by it self has little value to farmer. However, when combined with a yield monitor or a VRT, it may have value.

— Returns may be positive if costs can be spread over many applications. Specialized equipment, which has limited uses, has greater risks associated with it than equipment that has many uses. A multi-use tractor will likely pay for itself sooner than a new, single-use machine.

— Precision agriculture may not return on low-valued commodities as it does on high-valued specialty crops. Increasing yield by 5% through VRT may translate into $20/acre more revenue on a corn crop but $200/acre on wine grapes. The increased grape yield is more likely to pay for the PA system than the increased corn.

— GPS controlled tractor guidance systems may affect when and how tractors are operated.

Examples of Economic Analysis

Case 1. Evaluating Variable Rate Application

This case will evaluate the economics of variable rate application (VRA) of nitrogen fertilizer. In reality, soils can be tested for up to fourteen nutrients and some VRA systems can apply up to seven nutrients in one pass. However, only one nutrient is evaluated in this case in order to explain the evaluation process.

Suppose you have a field that has two distinct types of soils. The Low Yield soil can produce a maximum of 150 bushels of corn per acre, while the High Yield soil can produce 200 bushels of corn per acre. The Low Yield field can manage up to 200 pounds of nitrogen per acre. Applying more than 200 pounds will not affect the yield, up to some point. The High Yield field can utilize up to 250 pounds of nitrogen per acre. Again, applying more will not affect the yield, up to some point, but applying less lowers the yield. The field is approximately 50/50 High/Low Yield soils.

We will use Partial Budgets to evaluate four different scenarios.

Scenario 1: If you take only one soil sample in this field, you have a 50%/50% chance of testing the Low Yield or High Yield type of soil. If your test sample happens to be from a Low Yield area, you apply 200 pounds of nitrogen per acre, which is the "right" amount of nitrogen on the Low Yield area and not enough on the High Yield area.

Type of Soil	Max. Potential Yield (bu)	Fertilizer Required (pounds)	Fertilizer Applied	Actual Yield	Yield "Lost"	Fertilizer "Lost"
Low Yield	150	200	200	150	0	0
High Yield	200	250	200	150	50	0

Scenario 2: You take only one soil sample, which happens to be from a High Yield area. You apply 250 pounds of nitrogen per acre, which is "right" for the High Yield area and too much for the Low Yield area.

Type of Soil	Max. Potential Yield (bu)	Fertilizer Required (pounds)	Fertilizer Applied	Actual Yield	Yield "Lost"	Fertilizer "Lost"
Low Yield	150	200	250	150	0	50
High Yield	200	250	250	200	0	0

Scenario 3: You take two soil samples, one from a Low Yield area and one from a High Yield area and decide to average them. Hence, "on average" you should apply 225 pounds of nitrogen per acre, which is too much for the Low Yield area and not enough for the High Yield area.

Type of Soil	Max. Potential Yield (bu)	Fertilizer Required (pounds)	Fertilizer Applied	Actual Yield	Yield "Lost"	Fertilizer "Lost"
Low Yield	150	200	225	150	0	25
High Yield	200	250	225	175	25	0

Scenario 4: You take enough soil samples to determine which parts of your field are Low Yield areas and which are High Yield area and use VRT to apply the appropriate amount of nitrogen to each part of your field.

Type of Soil	*Max. Potential Yield (bu)*	*Fertilizer Required (pounds)*	*Fertilizer Applied*	*Actual Yield*	*Yield "Lost"*	*Fertilizer "Lost"*
Low Yield	150	200	200	150	0	0
High Yield	200	250	250	200	0	0

If you were concerned only with maximizing your yields, then you would of course prefer Scenario 4, where you take enough soil samples to determine the optimum amount of fertilizer for each area of your field, and then apply that amount. However, you are more likely concerned with maximizing your profits. As such, you want to determine if the added returns from more soil samples and VRT are greater than their costs.

Case 2: Grid-sampling vs. Representative Sampling

There are two distinct areas in your field is data. Your job as a manager is to edit and screen that data into information that affects your actions or prior beliefs, and hence has value. How you transform that data into information is in part dependent on how you obtained the data. Four situations obtained are presented below.

Soil samples done on a 1-5 acre grid

If the two areas are distinct soil types that respond to fertilizer differently and have different yields, then the data from the soil samples can be used for VRA. Your decision framework is similar to the Partial Budget example previously presented; i.e., will the expected returns be greater than the expected costs?

Soil samples are thought to be accurate for about three years. After that time, the soil should be sampled again to monitor any changes in nutrient requirements. Your decision is whether you should repeat your soil samples on a 5-acre grid; or reduce your costs by taking only two or three samples from the representative areas? The trade off is between the reduced costs of taking representative samples versus the reduced accuracy of representative samples. For a 1000 acre field, the comparable costs are:

Method	*Costs*
5-acre grid	$6/acre x 1,000 acres = $6,000
Representative samples	$18/sample x 3 samples = 54

The savings are large, but there is a risk with representative samples in that you will not know if the reduced accuracy affects yields until after harvest. Given the magnitude of savings and the possible loss from a wrong decision, additional data and information may be helpful

#2. A yield monitor linked to a GPS unit record areas of different yields

Having different yields in two distinctly different areas is a clear indication that the field can benefit from VRT rather than uniform treatment, but knowing that areas A and B have different yields does not say why the differences exist. Data from a yield monitor along with a soil map is a strong indication that the field has two types of soils and that VRA of nutrients is appropriate. However, without the soils map, the difference in yields could be due to a number of factors. Even with a soils map, it is possible that the difference in yield is due more to soil tilth and moisture retention characteristics than to nutrients.

#3. Veris system identified areas of different electrical conductivity

Electrical conductivity (EC) can be an indicator of a soil's water retention characteristics. If the difference in yields in areas A and B is due to water retention, then tests of EC may be an inexpensive way to obtain more data and hence more information about how to best manage the field.

#4. A near-infrared aerial photograph identifies areas of distinctly different foliage

Remote sensing in the form of a picture is data on the field. A physical investigation is required to transform that data into information about why the different foliage is there. Given the information as to the cause, the manager then may be able to do something to minimize damage.

The remote sensing by itself does not provide information nor provide value, but if the remote sensing provides data for the early detection of problems, it can minimize loss from damage. The economics of remote sensing and early detection of damage is difficult. The framework is simple:

— If the value of the crop x probability of loss > cost of detection,

— Then detection is economical.

It is similar to insurance: there is a small probability of a large loss, the expected value of which is larger than the premium charged. Some years there will be no indication of damage, hence no action needs to be taken. In those years the remote sensing has no value. Other years, however, the remote sensing will indicate damage in a timely manner, action will be taken to mitigate the damage, and the remote sensing had value by minimizing a potentially large loss.

Serendipities

Serendipities are unexpected benefits from some action or discovery. A classic serendipity was the development of the light emitting diode (LEM). To showcase the LEM, Texas Instruments threw together hand-held calculators for a trade show. The purpose was to demonstrate the LEM; the calculator was an after thought. The rest is history.

Many benefits from PA will be serendipitous discoveries. The engineers will develop the technology with one purpose in mind, and an enterprising farmer will discover a much better use, in terms of economic benefits. Yield monitors are one such example. A yield monitor with GPS was developed to provide detailed maps of variation in yield across a field. An unexpected benefit is that farmers with yield monitors now know within a pound how much they have loaded onto a truck. Both grain and tomato growers in California have reported virtually eliminating the probability of being stopped for an overloaded truck because of their yield monitors. Some farmers have been buying yield monitors without GPS, strictly to have accurate measures of their truck loads.

12

Small Sector Precision Farming

Precision farming provides a new solution using a systems approach for today's agricultural issues such as the need to balance productivity with environmental concerns. It is based on advanced information technology. It includes describing and modelling variation in soils and plant species, and integrating agricultural practices to meet site-specific requirements. It aims at increased economic returns, as well as at reducing the energy input and the environmental impact of agriculture.

"Precision Farming" or "Precision Agriculture" is capturing the imagination of many people concerned with the production of food, feed, and fiber. It offers the promise of increasing productivity, while decreasing production costs and minimising the environmental impact of farming. There are three fundamental elements in this technology (Fig. 1).

Describing variability is the key concept. In particular, it is based on variation within each field. Variability should be understood in at least three aspects: spatial, temporal and predictive.

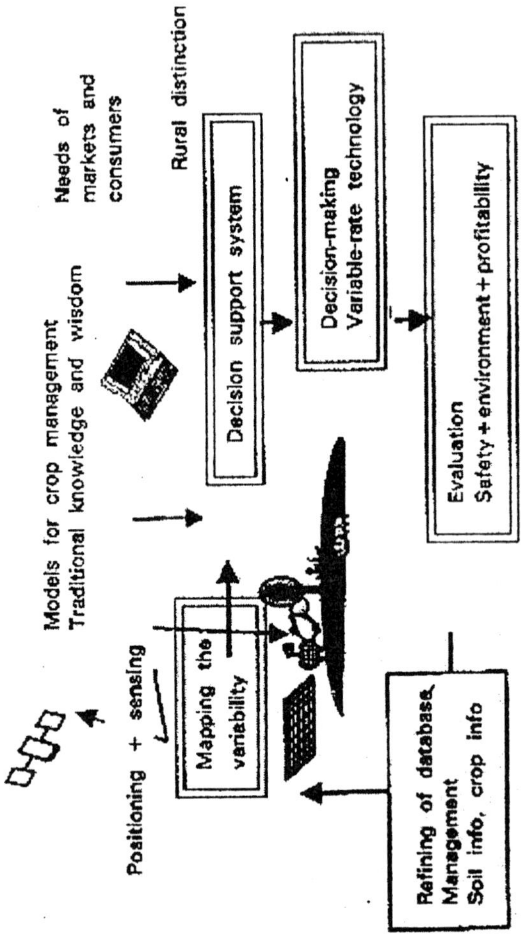

Figure 1. How precision farming works

Variable-rate technology (VRT) is used to adjust the agricultural inputs according to the site-specific requirements in each part of the field. If machines are used, this requires variable-rate machinery. On small farms, inputs can be applied manually. Variable-rate applications need:

— Correct positioning in the field;

— Correct information at the location; *and*

— Timely operations at the site concerned.

Decision support systems offer a range of choices to farmers with respect to trade-off problems where conflicting demands must be taken into account, such as productivity and protection of the environment. This approach helps to optimise the whole farming system.

Information Levels

There are four levels or stages in the quality of information. The lowest level is data, followed by information, knowledge, and finally wisdom (Fig. 2). The "data-stage" means a mass of signals and numerical values, which have no practical value in themselves. The "information-stage" provides some meaning from a set of data, such as levels of excessive, appropriate or deficient fertiliser use. The "knowledge-stage" implies that the information is individualised in some logical way which can enable someone to make a decision, such as application guidelines. The "wisdom-stage" belongs to an individual who finds an original solution, such as the approach of the skilled farmer.

Information technology tends to be powerful in levels up to the knowledge-stage. The wisdom-stage requires the intellectual and creative activities of farmers and researchers, if there is to be a break-through in accumulated knowledge. Precision farming needs all stages of information in the agricultural production system, and also requires good linkage between the stages. In particular, information technology should be closely linked to farmers.

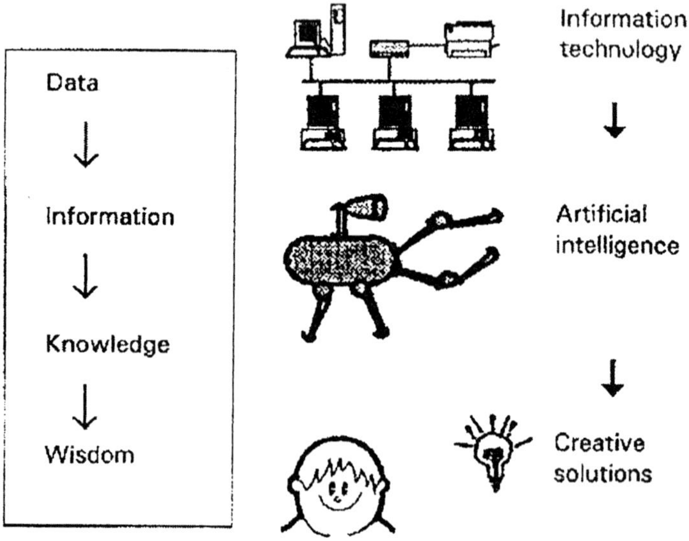

Figure 2. Level of information

Developing system technology for precision farming is shown in Fig. 3. First of all, we need to describe and understand the variability within and between fields. Field sensors with GPS and monitors for machine application make this easier. The next stage is to develop machines which can be operated by remote control.

There are three steps in technology development, and three strategies for precision farming, as shown in Fig. 4. Step 1 is based on conventional farming technology, with intensive mechanisation to reduce the labour input. Step 2 involves the development of mapping techniques, VRT

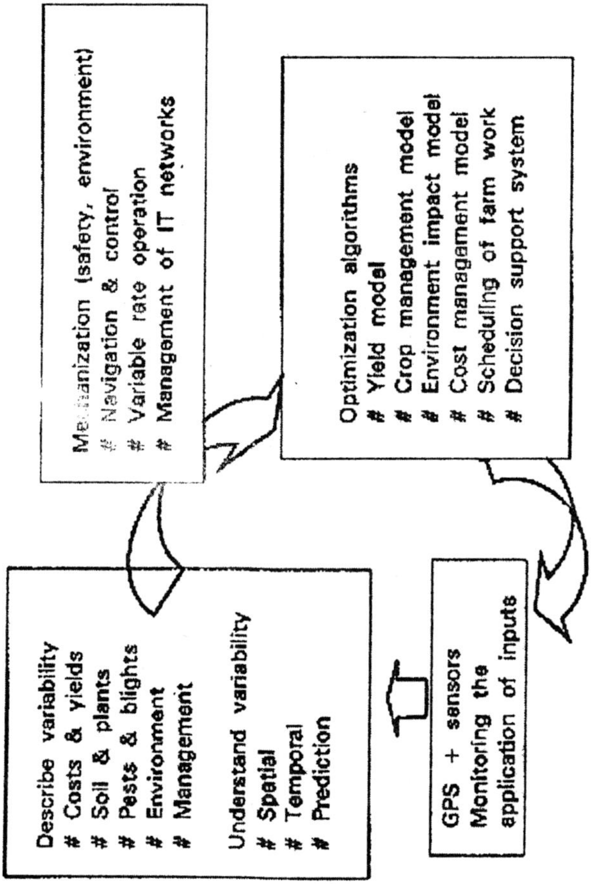

Figure 3. Development of precision farming technologies

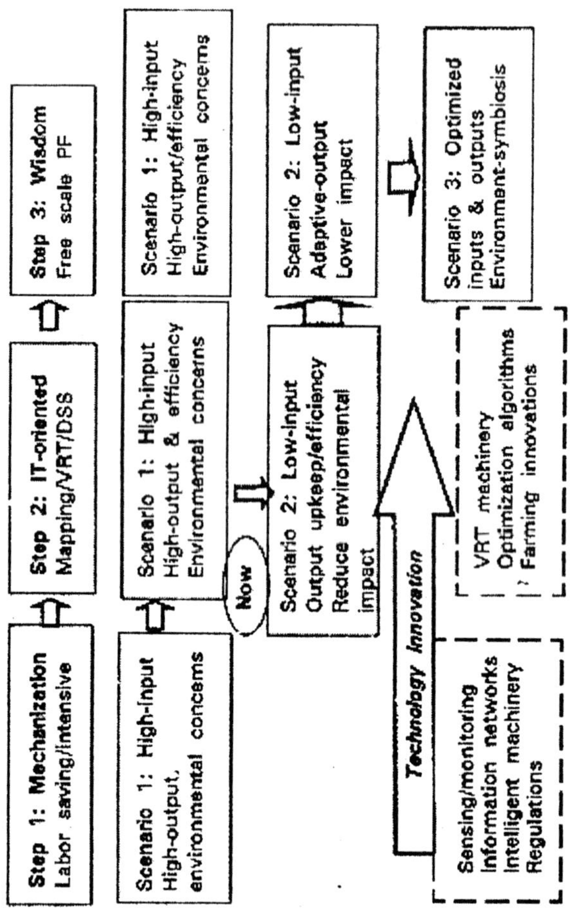

Figure 4. Development of technology level and farming strategies

machines, and introductory DSS on the basis of information technology. Step 3 implies the maturity of wisdom-oriented technologies.

Scenario 1 is based on a "high-input and high-output" conventional strategy. Scenario 2 has a strategy for "low-input but constant-output", and Scenario 3 aims at "optimised input-output" as the goal of precision farming. Advanced technology levels allow us to choose freely between these three scenarios. Effective regulations will encourage progress in precision farming.

In Japan, technology at the moment is at the level between Step 1 and 2, while the farming strategy is changing from Scenario 1 to 2. The shift from Step 2 to 3 involves a drastic change in the farming system. In the mature stage of Step 2, each field has information added which makes possible the best management of that field. In Step 3, all factors of the farming system are well-organised for precision farming. This allows us to manage regional variability, as well as the local variability within a single field.

In general, a farming system involves five factors. These are: plant variety, field features, technology, regional infrastructure, and the motivation/intentions of the farmer. Better integration of these five factors can create a competitive farming system which suits local conditions.

Precision farming uses field maps, variable-rate technologies and a decision support system. As shown in Fig. 5, generating the field maps is in itself an important source of information. Variable-rate technology not only increases productivity by re-organising the three factors of technology, plants and fields, but also creates a better linkage with the regional infrastructure, e.g. by following environmental regulations. A decision support system provides the best technology, taking into account the aims and motivation of farmers as well as environmental factors. In other words, precision farming brings about an innovation in the whole system of agriculture.

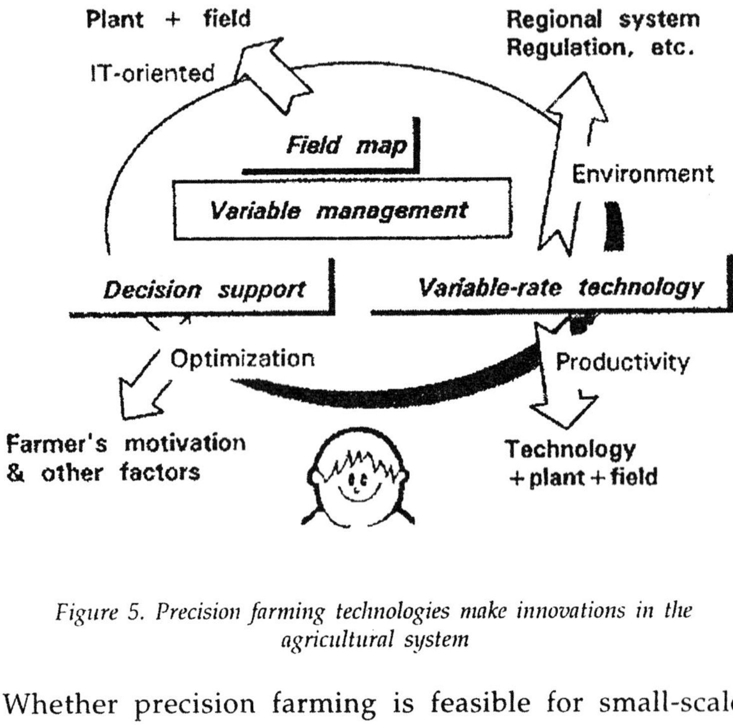

Figure 5. Precision farming technologies make innovations in the agricultural system

Whether precision farming is feasible for small-scale farms is a leading issue for agricultural scientists and politicians in Japan. It should be noted that precision farming is characterised by variable management. A key point in precision farming is understanding variability in the field. There are at least two types of variability. One is within-field variability, the other is between-field or regional variability.

Within-field variability focuses on a single field, and the one plant variety being cultivated. Between-field variability considers each field as a unit on a map. We need to consider what kind of variability is involved when we consider precision farming for small farms. Whether farms are large or small, precision farming should mean improved farm management. It should give a higher economic return with a reduced environmental impact.

On a single small farm, the farmer can understand fairly well what is going on in each field. This makes possible variable-rate applications to meet site-specific requirements, using the farmer's knowledge and skills. When it comes to an area of a few dozen hectares, containing many small fields, precision farming has to coordinate diverse types of land use and many farmers with different motivations, as shown in Fig. 6.

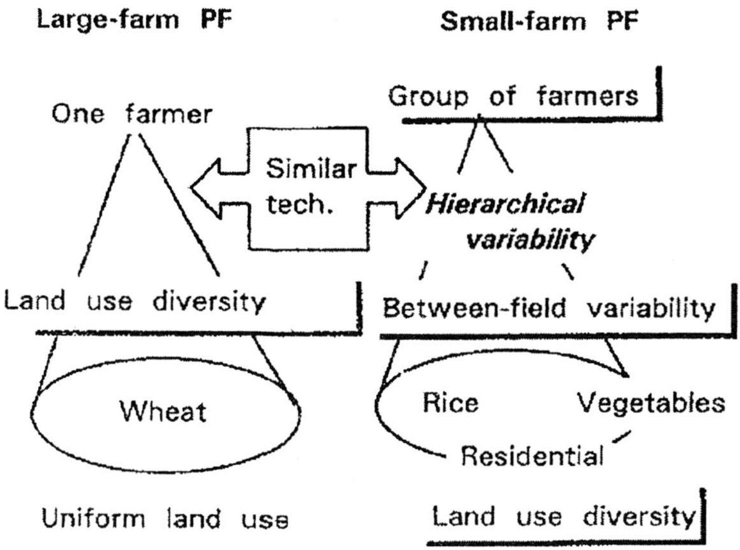

Figure 6. Regional precision farming (PF) for small farms

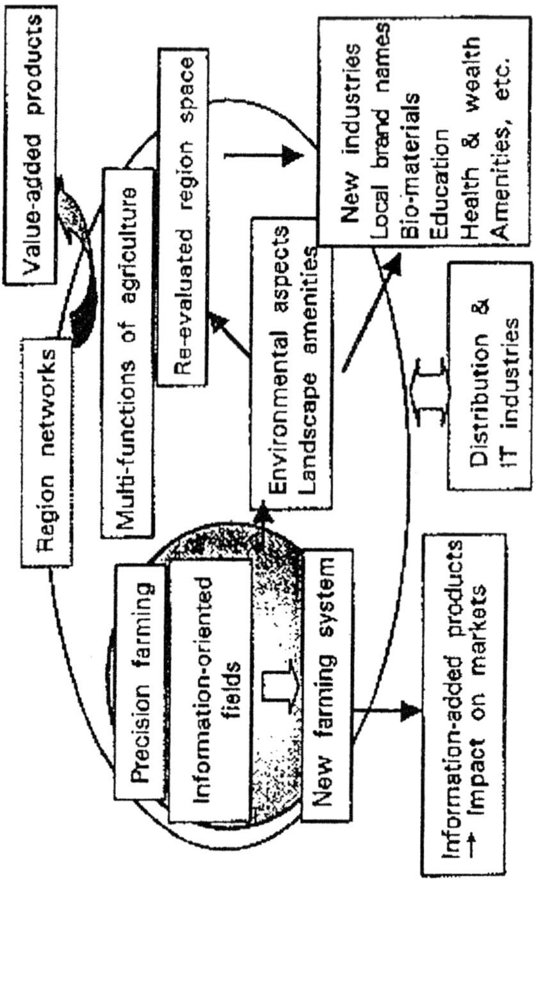

Figure 7. Role of precision farming in regional development

Regional precision farming must manage a hierarchy of variability: within-field, between-field and between-farmers. High-tech approaches, such as a yield meter with GPS, are available for regional precision farming covering many small farms.

Moreover, measures to conserve or improve the environment should be undertaken on a similar scale. From the point of view of development in a rural area which includes small farms and local companies, precision farming offers the possibility of developing a new kind of industry, by fusing agriculture to various kinds of industrial activity (Fig. 7). If the multi-functions of agriculture are re-evaluated using information-added fields, value-added space of this kind can be seen as providing new resources, such as new biological materials, open-air classrooms and green tourism.

Real-time Soil Sensing

In Japan, paddy fields occupy half of the agricultural area, producing about ten million metric tons of grain every year. Paddy rice production is a kind of hydroponic system, with well-organised irrigation and drainage facilities. In Japan, it is also highly mechanised.

Paddy rice production is very productive, but environmental concerns have become national issues in Japan. The problem is how to manage precisely the paddy production system while considering the environmental impact. Solving this problem requires a good understanding of what is happening in the paddy field.

Shonk *et al.* developed a portable soil organic matter sensor with photodiodes using a single wavelength. It gave good results in predicting soil organic matter in the range 1.5 - 6%. Sudduth and Hummel investigated the feasibility of spectral reflectance to sense soil organic matter. A portable NIR spectrophotometer was designed to evaluate soil organic matter, CEC and moisture content in a ploughed soil at a depth of 3.5 - 5 cm. This approach can be useful to get information about the field surface, but we still need *in situ* soil sensing in the zone of root development for practical use in crop management. Shibusawa *et al.* have developed a real-time soil spectrophotometer with an RTK-GPS to sense underground soil parameters at depths of 15 - 40 cm.

Soil Spectrophotometer

The soil spectrophotometer used is shown in Fig. 8.

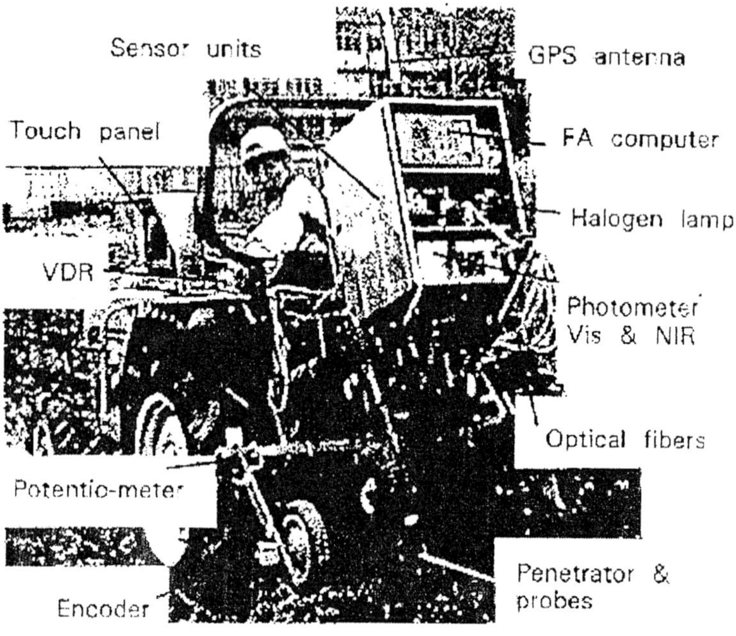

Fig. 8 Revised soil spectrophotometer

It was designed to collect data on soil reflectance at depths of 15 to 40 cm. The sensor system was composed of three main units: the external housing, the soil penetrator and probes, and the external sensing and monitoring devices.

The penetrator tip with its flat edge cuts the soil in a uniform way. The plane edge behind it smooths the soil to produce a uniform surface. Inside the housing are seven micro optical devices. Two optical fiber probes, using light energy at wave-lengths of 400 - 2400 nm, are used for illumination, giving an illuminated area of about 50 mm diameter on the soil surface. Two additional optical fiber probes are used to collect soil reflectance in the visible and NIR ranges. One fiber bundle passes reflected energy in the 400 - 900 nm wavelength range, while the other optical probe carries reflected energy in the 900 - 2400 nm wavelength range. A micro CCD camera is adjusted to monitor a 75-mm focus point on the soil surface.

The sensor unit's housing includes the core devices of the system, such as a 150 W halogen lamp, a spectrophotometer, a FA computer (IBM, PC/AT, Pentium MMX), a RTK-GPS (Trimble MS740) receiver, etc. The spectrophotometer has a 256-channel linear photodiode array to quantify the reflected energy in the 400 - 900 nm wavelength range. A 128 channel linear photodiode array is used to quantify the reflected energy in the 900 - 1700 nm wavelength range. Data scanning time is just over four microseconds. Integration of scanned data is carried out for each individual scan to get average values.

A video data recorder on the tractor displays images of the soil surface during the experiment. The displayed images are used to monitor operations in case of emergency, such as blockages or obstacles. The images also provide information about which data should be omitted from data analysis. The liquid crystal monitor serves as a touch control panel, and a mouse and keyboard are also available for accessing the FA computer.

Field Tests and Results

The experiment was conducted in a 0.5 ha paddy field on the Experimental Farm of Kyoto University, Japan in December 1999. The soil texture of the fields was 47% sand, 30% silt and 23% clay. The working speed was about one kilometer per hour. Scans were at approximately one-meter intervals. It took approximately 20 microseconds to do the scan, and three seconds to record the data. The spectrum data was collected at intervals of about 5 m, which gave more than 800 locations for soil reflectance data. The working depth was 200 - 250 mm.

For calibration purposes, 25 soil samples were collected at the same location and depth as the scanning points and analysed in the laboratory for moisture, organic matter content, nitrate (NO_3-N), pH and EC. Fifteen samples were used for calibration, and the remaining ten samples were used for validation. A standard moisture content was obtained, by keeping the samples 24 hours in an oven at 110°C. Soil organic matter content was evaluated as the loss after four hours of combustion in an oven at 800°C. NO_3-N, pH and EC were analysed in the clear layer at the top of muddy water, using portable ion meters. The test muddy water was provided by diluting 5 g dried soil with 25 g distilled water, stirring the mixture for 30 minutes, and then leaving it to stand for 24 hours.

Table 1. Results of calibration and validation for soil parameter prediction

Parameters	Treatment	Calibration (n = 15)		Validation (n – 10)		Wavelength (nm)
		R^2_{cal}	SEC	R^2_{val}	SEP	
Moisture (% db)	2nd deriv.	0.908	1.893	0.655	3.111	606,1329,1499
SOM (%)	2nd deriv.	0.948	0.259	0.647	0.559	806,1311,1238
NO_3-N(mg/100g)	2nd deriv.	0.803	3.699	0.539	4.741	589,1014
pH	1st deriv.	0.714	0.101	0.541	0.127	957,1214
EC(μS/cm)	2nd deriv.	0.735	23.975	0.650	41.718	456,984,1014

Table 2 Summary of semivariance analysis

		Semivariance factors					
		Nugget	Sill	Range	Proportion	R2	RSS
Parameters	Model fit	Co	Co + C	Ao or 3Ao	C/(Co + C)		
Moisture	Spherical	16.20	47.64	38.60	0.66	0.96	34.74
SOM	Spherical	0.58	1.19	29.20	0.51	0.93	0.01
NO3-N	Spherical	47.50	136.20	34.50	0.65	0.94	302.40
pH	Spherical	0.06	0.174	40.40	0.67	0.99	1.29E-04
EC	Spherical	2474	5562	46.60	0.56	0.93	586760

For the spectral reflectance, four stages were followed. The first stage was linearisation with a Kubelka-Munk transform, while the second stage was elimination of optical interference in the spectral data with a multiplicative scatter correction. The third stage was to reduce the number of wavelengths used for calibration with correlation analysis, including derivative operation. The final stage was the calibration stage using the stepwise multiple linear regression analysis with S-Plus Data Analysis Software. The calibration model was quantified using the standard error for calibration (SEC), standard error for prediction (SEP) and coefficient of determination (R^2).

Semivariance analysis was performed using the GS+ Geostatistics Software, and soil maps were obtained by the block kriging method. Results of calibration and validation analysis (Table 1) produced higher scores of R^2 and fewer errors for the respective soil parameters. The second derivatives of light absorption tended to provide best-fit prediction models.

With the prediction models, values for soil parameters were evaluated at 860 locations in the field. The means and standard deviations were 48.4% and 6.5% for moisture content, 9.51% and 1.06% for organic matter (OM) content, 42.1 mg/100g and 11.0 mg/100g for NO_3-N content, 6.83 and 0.39 for pH, and 173.1 µS cm and 69.6 µS/cm for EC. Based on these values, semivariance analysis (Table 2) was performed. Within the experimental field, the soil OM content had the lowest spatial correlation (29.20 m), followed by the NO_3-N content (34.50 m), moisture content (38.60 m), the pH (40.40 m), and the EC (46.60 m).

13

Comprehensive Precision Farming

Precision farming is a comprehensive approach to farm management and has the following goals and outcomes: increased profitability and sustainability, improved product quality, effective and efficient pest management, energy, water and soil conservation, and surface and ground water protection.

In PF, the farm field is broken into "management zones" based on soil pH, yield rates, pest infestation, and other factors that affect crop production. Management decisions are based on the requirements of each zone and PF tools (e.g. GPS/GIS) are used to control zone inputs. In contrast, traditional farming methods have used a "whole field" approach where the field is treated as a homogeneous area.

Decisions are based on field averages and inputs are applied uniformly across a field in traditional farming. The advantage of PF is that management zones with a higher potential for economic return receive more inputs, if needed, than less productive areas. Therefore, the maximum economic return can be achieved for each input.

Information, Technology, and Decision Support

PF relies on three main elements: information, technology, and decision support (management).

Timely and accurate information is the modern farmer's most valuable resource. This information should include data on crop characteristics, hybrid responses, soil properties, fertility requirements, weather predictions, weed and pest populations, plant growth responses, harvest yield, post harvest processing, and marketing projections. Precisionfarmers must find, analyse, and use the available information (Figure 1) at each step in the crop system. An enormous database is available on the internet. This data is both accessible and quickly updated.

Figure 1. The learning cycle is on-going, and precision farming tools provide a means to record and save information for year-to-year comparisons of a location.

Precision farmers must assess how new technologies can be adapted to their operations. For example, the personal computer (PC) can be used to effectively organise, analyse, and manage data. Record keeping is easy on a PC and information from past years can be easily accessed. Computer software including spreadsheets, databases, geographic information systems (GIS), and other types of application software are readily available and most are easy to use.

Another technology that precision farmers use is the global positioning system (GPS). GPS allows producers and agricultural consultants to locate specific field positions within a few feet of accuracy. As a result, numerous observations and measurements can be taken at a specific position. Global information systems (GIS) can be used to create field maps based on GPS data to record and assess the impact of farm management decisions. Data sensors used to monitor soil properties, crop stress, growth conditions, yields, or post harvest processing are either available or under development. These sensors provide the precision farmer with instant (real-time) information that can be used to adjust or control operational inputs.

Precision farming uses three general technologies or sets of tools:

— Crop, soil, and positioning sensors - these include both remote and vehicle-mounted, "on-the-go" sensors that detect soil texture, soil moisture levels, crop stress, and disease and weed infestations;

— Machine controls - these are used to guide field equipment and can vary the rate, mix, and location of water, seeds, nutrients, or chemical applications;

— Computer-based systems - these include GIS maps and databases that use sensor information to "prescribe" specific machine controls.

Decision support combines traditional management skills with PF tools to help precision farmers make the best management choices or "prescriptions" for their crop production system. Unfortunately, decision support has many times been either unreliable or difficult to understand. Building databases based on the relationships between input and potential yields, refining analytical tools, and increasing agronomic knowledge at the local level are yet to be accomplished. Most agricultural researchers agree that decision support remains the least developed area of PF. Diagnostic and database development will eventually replace technologies as the real benefit of PF.

Getting Started

PF is not appropriate for every field. To determine if a specific field will benefit from PF, use the following steps:

Review current data. Review existing information such as soil survey maps, cropping practice records, historical characteristics, and hand-drawn maps that show weed and disease problems, wet areas, and other field characteristics. Use existing maps or aerial photographs and draw boundaries around areas that have different characteristics (e.g. soil type, crop yield, or soil pH).

Obtain additional data. Several government agencies have produced maps including the following:

- — digitised soil surveys (SSURGO)
- — digital line graphs (DLG)
- — digital ortho-photo quarter quads (DOQQ)
- — Topologically Integrated Geographic Encoding and Reference (Tiger) files, which describe roads and boundaries

Each of these maps contains unique data for a specific purpose (e.g. roads, soil type, contour). Most information can be acquired at little or no cost.

Obtain recent aerial photographs of your farm. Photographs taken when fields were barren and at two or three vegetative stages are especially useful. These should be available from your local office of the Farm Service Agency (FSA). Use the photographs to identify known property characteristics such as fence lines, sandy areas, poor drainage areas, livestock usage, and other features.

Gather yield data. Determine the yield variation within each field by using a yield monitor. Many yield monitors can be retrofitted to your combine. Sometimes, the purchase of a yield monitor is too expensive or a yield monitor does not exist for a particular crop. In these cases, you can obtain a rough yield estimate by spot-checking various areas of the field with a weigh wagon.

Examine results. Mark areas with different yield values on an aerial photograph. Colour-coding with a felt marker or crayon makes referencing easier. For example, use green for areas with high yields, yellow for areas with medium yields, and red for areas with low yield. Compare yield data with physical characteristics of the field and field notes. If you notice a pattern of cause and effect, such as a very heavy weed infestation where yields are reduced, you probably need to increase your management of that area. You may wish to collect soil samples across the field to determine if soil nutrient levels limited the yield in less productive areas. Keep a notebook for each field that contains notes, photographs, and maps of each year's harvest.

Data interpretation. You may notice patterns of uniform and non-uniform variability throughout the field when interpreting yield maps. Table 1 provides a guide to interpreting variability within a yield map. Use this information to evaluate management techniques and other influences on crop production. Verify the reasons for non-uniformity found in yield maps by on-site inspection.

Table 1. Guide to interpreting (or detecting) variability within a yield map (or field). Visual observations from a yield map can be seen as having uniform or irregular patterns

Pattern Description/Explanation Uniform Patterns With Direction of Application
— change in planting date — change in hybrid/variety — change in chemical application — selected rescue treatment — chemical skips and misapplications — equipment errors — poor straw/chaff distribution — compaction
Against Direction of Application
— drain tile patterns — historically different fields — old traffic patterns — manure applications — pipelines/phone lines underground — previous compaction
Irregular Patterns Irregular Line
— topography changes — herbicide drift — border shading effects — insect infestation from bordering lands — improper manure application — waterways
Irregular Area/Patch

— change in soil type
— drainage patterns
— weed infestations
— soil fertility changes
— previous crop activity
— disease infestations
— herbicide carryover
— historic occurrences
— insect infestations
— changes in organic matter
— animal damage
— wet areas

Use a systematic approach to storing information when collecting data. Make paper or electronic copies of your information. Store this information in different places from your usual record keeping area to prevent theft or damage from fire, water, or some other disaster.

Computers with PF and record keeping software help manage and retrieve data. GIS programs overlay geo-referenced information for further analyses. You may find working with a computer difficult at first. If so, use services that collect, handle, analyse, and interpret data for you.

Management strategy. Determine a course of action once a problem has been identified. This is sometimes difficult since each farm is unique and a prescribed solution may not be available. Use the services of an agricultural consultant, county Extension agent, or Extension specialist to evaluate your management strategy.

Reassessment

Precision farmers must continuously reassess the profitability of their PF program. This can be done by comparing yield monitor data to the financial records for a specific field. If a decision based on PF did not improve profitability, a more or less intensive PF management scheme may or may not be warranted.

Examples of Site-Specific Crop Management

Developing management strategies is the most difficult part of the PF process. Precision farmers must consider the steps of crop production that can best be controlled and managed to provide the greatest economic return. The following examples provide a framework for developing a PF management plan. Although conceptual in nature, these examples are common occurrences on farms and could be used with any management strategy.

Example 1. Nutrient Management Plan.

A farmer wants to "fine tune" his nutrient management plans for optimal economic returns. He starts by evaluating the soil characteristics (the lower right corner of Figure 2) within a field. Next, the farmer classifies the field into areas or "management zones" based on common soil types or crop potentials. Within these zones, the farmer decides on the combination of operating strategies (e.g. conservation tillage with variable rate fertiliser application) needed to obtain the maximum profit from that field. He can fine-tune the nutrient management plan based on weather and market forecasts.

The fine-tuning job does not end at harvest. The farmer should gather yield data and information on the amount of residual nutrients not utilised by the crop. He should compare these results to his yield goals and economic returns to determine whether his management decision worked.

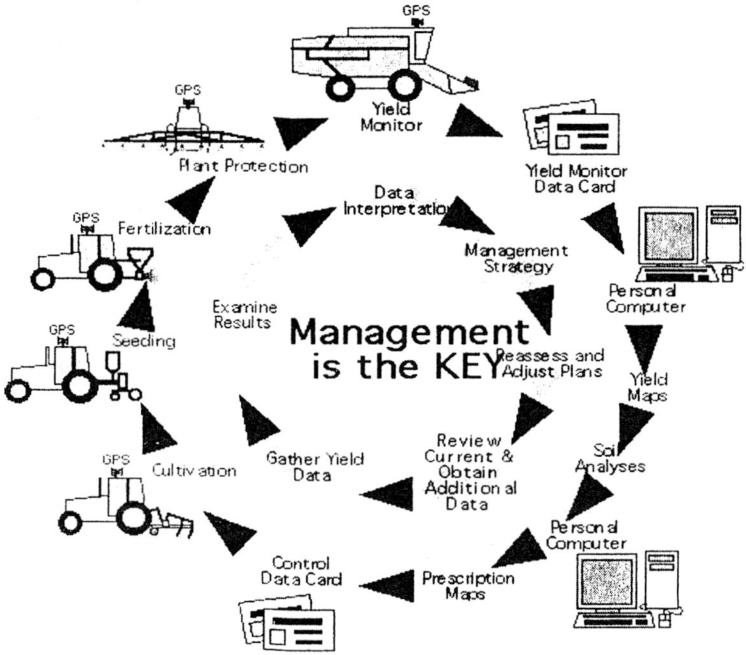

Figure 2. The learning cycle with equipment use and technology overlaid

Example 2. Management Strategy to Reduce Chemical Costs

A farmer wants to save money by applying pesticides to the correct target at the proper rate and time. The farmer selects a high accuracy applicator equipped with GPS-based controllers. Next, he obtains field scouting or remote sensing data of pest-infestation zones within a field and develops a geo-referenced map. The services of an integrated pest management (IPM) specialist may be used to determine the optimal time for pesticide application.

The farmer sprays only those areas that have both a high potential economic return and a high threshold of pest infestation. The GPS-based controls ensure that those areas receive the proper application rate. The controller also logs actual rate, time, and location data. As in Example 1, the farmer should compare his harvest to his yield goals and economic returns to determine whether his management decision worked.

A farmer must continually ask questions to optimise his PF program. These may include:

— Is there error in the data collection process?
— How accurate is the information?
— What does the information tell me?
— Do any observations indicate variation from other influences?
— Is one year of information sufficient? (not likely)
— Can I justify making final management decisions based on this information?
— Do I need additional information?
— How much time and money have I spent on collecting information and making decisions?
— Have the decisions to correct problems influenced succeeding year's yield map?

14

Public Involvement in Precision Agriculture

Precision agriculture developed by assembling the components of technologies developed for purposes far removed from agricultural production. The private sector made a significant investment to tailor information technologies for agricultural applications, but the public sector agricultural research community contributed relatively little. It is anticipated that private sector investments in development and diffusion of precision agriculture will continue at a rapid pace. The array of possible public sector roles ranges from research, education, and extension, to development of infrastructure for communications and institutional underpinnings for intellectual property rights.

The public sector has primarily contributed to precision agriculture indirectly through large infrastructure investments outside of agriculture. The largest and most critical investment has been in GPS development and implementation, which was motivated by U.S. Department of Defense (DOD) needs for accurate and instantaneous navigational positioning across the world. The $12 billion invested since 1990 largely overlooked potential civilian spin-off applications, including those in agriculture.

As adoption of precision agriculture increases, explicit public policies could be formulated to foster or retard adoption. These should be focused on public benefits from adoption that do not compete with private industry objectives and cannot be realised exclusively by any one individual or company. An important reason for public involvement is to avoid any unintended consequences and dangers that might be caused by the increasingly widespread conversion to precision agriculture technologies. The following ways that public involvement could be justified to further appropriate development and dissemination of precision agriculture already undertaken in the private sector:

- — providing information on advantages and disadvantages of precision agriculture to potential adopters through research, development, and technology transfer and dissemination activities;
- — providing education and human capital infrastructure;
- — stimulating public data collection efforts;
- — providing physical infrastructure in cases where there are substantial economies of scale; and
- — protecting private property rights, particularly with respect to intellectual property such as software and data ownership.

Role of Research and Development

The public sector in the United States has played a major role in research, development, and dissemination of new agricultural technologies for more than a century. Until the onset of World War II, agricultural research at USDA was the principal component of federal research and development efforts, accounting for almost 40 percent of all federal R&D spending. In 1991, by contrast, USDA research spending accounted for only 2 percent of federal research and development spending. Public sector agricultural research and development has changed little in real terms since 1980. Agricultural research at the state level has been carried out by state agricultural experiment stations (SAESs) under the direction of the land grant university system created by the Morrill Act in 1862.

Today the public sector apparatus for research and development of new agricultural technologies consists of USDA's research arms (Agricultural Research Service, Economic Research Service, and National Agricultural Statistics Service) and the network of SAESs. USDA helps set SAES research and development priorities through competitive grant funding overseen by CSREES. In 1992 total public research and development spending on agriculture was about $2.9 billion, or 46 percent of public and private agricultural research and development spending in the United States.

Until 1978, real expenditures on agricultural research and development from private sources equalled public expenditures. Since then, public funding has remained nearly flat at about $2.5 to 3 billion, while private investments have increased to nearly $4 billion. The totals mask a significant shift in emphasis in the type of agricultural research conducted by the private sector. In 1960, more than 80 percent of private research funding was for improving farm machinery or developing new food products or processing methods, while public research focused on increasing crop and livestock yields. By 1992, 60 percent of private research was also devoted to increasing crop and livestock yields by improving crop varieties, agricultural chemicals, animal breeding, feeds, and pharmaceuticals. These trends point toward more potential for competition between private and public agricultural research and development and a less clear-cut division of labour.

Most of the research and development embodied in current precision agriculture technologies has come about either through public investments in defense and space technologies or by the private sector; there has been little investment in precision agriculture by traditional public agricultural research institutions. There is every reason to believe that private research and development investment in precision agriculture will continue to be made as long as there is potential for profit. What is not clear are appropriate roles for public research and development in precision agriculture that are not duplicative of private efforts and that can materially improve development and adoption.

Left to itself, the private sector will generally underinvest in socially desirable research for several reasons:

- Gains from research investments may be difficult to protect from competitors.
- Basic research may be too risky to justify investment.
- Potential markets for products of research may be too small.
- Traditional technologies may have fully captured the market.
- Available labour is not trained to use the new technologies.
- Clients may have no incentives to adopt products of research, particularly those that improve environmental quality.

Research and development are costly, and it is difficult for firms to appropriate the fruits of their research and development efforts because, once known, the results of those efforts can be copied easily and inexpensively. For example, competitors of a firm that has invented a new piece of equipment can reverse-engineer their own versions and thus produce equipment of equal or better capability without having to invest in the initial research and development.

Patents and other forms of intellectual property rights were designed to encourage the private sector to conduct research. Patent law is geared toward protecting inventions that embody new knowledge, not toward protecting what Huffman and Evenson term pretechnology science (i.e., scientific research applied to specific problem areas but not toward the development of products, inventions, or other patentable items), which is thus generally neglected by the private sector. For example, research on farming methods enhances knowledge about crop productivity under alternative management systems.

Such pretechnology science cannot be patented, and the private sector has an incentive to engage in research of this kind only if the resulting knowledge is expected to increase sales of a particular product, such as a specific agricultural chemical or piece of farm equipment. At the same time, knowledge of this kind is too specific to be of interest from a general scientific point of view and thus tends to be neglected by purely academic researchers as well. Where the private sector does not have incentives to conduct such research, public sector research and development may be required to fill the gaps.

In some cases, research and development may be too risky for the private sector to undertake. Of particular importance are pilot inventions, the early prototypes of entirely new kinds of technology. For example, many crop breeding methods were developed at SAESs; private sector research on crop breeding became significant only after the introduction of hybrid seed technology provided a natural means for a firm to protect its investment, because producers were required to purchase new seed each year. In the pesticide industry, as well, innovations have come mainly from the private sector.

Chemical companies do share information and conform to standards on toxicities, dangers, and other factors, and there is a private/public partnership in education and training for pesticide use. The key to cooperation and investment by the private sector has been regulation, combined with public research and extension programs which have protected producers and the environment from misuse of pesticides.

Markets for new technologies may sometimes appear to be too small to permit private firms to recoup their research and development costs, even if the invention would benefit society as a whole. In such cases, investment in private research and development will be lacking. In the long term, most of the benefits of technological improvements in agriculture accrue to consumers; competition eventually limits both producers and equipment suppliers to returns on investment equal to the cost of capital.

Traditional or conventional technologies may have fully captured the market, making it difficult for new technologies to emerge until some event occurs to disrupt markets. The long, slow development of conservation tillage illustrates the difficulty of penetrating a market that is dominated by conventional methods. Energy shocks in the mid-and late-1970s caused a disruption that conservation tillage, because of its resulting energy savings, could exploit, and conservation compliance policies added a further incentive to change from traditional tillage.

Available producer and hired labour may not have sufficient training to use new technologies, limiting market potential. Lack of training in GIS and GPS electronics may limit adoption of these technologies and retard investment in them by the private sector. Some investments in making the systems easier to use, understand, and interface with computer and other systems may overcome an initial lack of training in the labour force.

Finally, the private sector similarly has little incentive to engage in research aimed at enhancing environmental quality (i.e., by reducing pollutant emissions), even though the results of such research may be of great value to society. Because agricultural pollutants are generally not regulated, and because environmental effects are by-products of production that do not show up on the bottom line, producers may be unwilling to pay for products embodying such research.

Technology development firms may thus have little or no incentive to engage in research to reduce environmental spillovers from agricultural activity. Increasing public concern for the environment may encourage technology providers and producers to adopt practices that enhance environmental quality, especially if they add little to production costs, but current legal or administrative requirements offer few direct incentives to do so.

Measurement Methods

The potential of precision agriculture is limited by the

lack of appropriate measurement and analysis techniques for agronomically important factors. Public sector support is needed for the advancement of data acquisition and analysis methods, including sensing technologies, sampling methods, database systems, and geospatial methods. Although many of the technologies making up precision agriculture are relatively mature (i.e., GPS, GIS, and remote sensing), there remains room for improvement in many technological areas directly related to agricultural applications.

One of the most important of these is the development of local sensors that can be used on farm equipment to determine crop stage, soil conditions and chemistry, weed concentrations, presence of insects, and other variables important for crop growth. Public sector researchers should concentrate on the basic scientific principles that could underlie new sensor development and on the relationships between measurements from such new sensors and modelling of crop growth and yield. Private sector research and development is more appropriate for making the new sensors operational and marketable.

Another farm management problem of special importance is the determination of optimal sampling strategies. Some precision agriculture technologies function by permitting adjustment of farming practices (i.e., input application rates) to match variability in production conditions, such as soil nutrient levels or other aspects of soil quality. Determining the extent of variability is essential, not only at the subfield level, but at all spatial levels. Optimal sampling depends on trade-offs between potential savings in input expenditures, potential gains from increased yields due to improved management, and sampling costs.

Database and GIS systems include interpolation algorithms to predict data at intermediate points, but no existing research validates assumed projections under true agronomic variability in the presence of obvious measurement errors. Geospatial methods must be advanced and incorporated into GIS to facilitate accurate analysis and inference from collected precision agriculture data. The public sector should take the lead role in researching (a) the nature of variability within farm fields and at other spatial scales, (b) the required precision of compatible measurements that are to be included within GIS data sets, and (c) the fundamental geospatial analysis methods necessary for interlayer correlation analysis and inference.

Both the public and private sector have been involved in developing and disseminating standards for hardware, software, and data interpretation that could influence precision agriculture development and adoption. Such standards have been critical in the development of general computer technology, and are emerging in GIS (such as the spatial data transfer standard and open GIS standards). Developing standards always involves a trade-off between ordering the chaos of individual systems and stifling creative breakthroughs in emerging technologies, and thus must be carefully managed.

From the perspective of the user, standardisation would facilitate data interchange, particularly moving spatial data from one proprietary software package to another and to regional databases. Hardware interoperability would facilitate connection of technologies and equipment into a unified system (i.e., a VRT controller in a tractor cab communicating with separate rate controllers for seeds, liquid, and dry chemicals). Standards affecting data and hardware interchange affect the integration and ease of using these new technologies. Precision agriculture is technically possible today, in large measure, but requires a high degree of technical know-how and persistence, much as did the early personal computer systems.

Precision agriculture developers and vendors are torn between conflicting goals of responding to user needs and maintaining proprietary advantages, market niches, and demand for system and electronic consulting services. There are also some potential conflicts between publicly provided services and private vendors. Most concerns in precision agriculture relate to spatial data standards, because many aspects of conventional database management and operating systems already have information technology standards developed by industry and public consortia. These include:

— government standards,

— consortium standards, and

— ad hoc or default standards (i.e., from dominance in the market, such as AutoCad DXF or Arc/Info Export).

Several paths of development and implementation could take place, with different trade-offs in timeliness, responsiveness, and enforceability.

In the arena of precision farming, the Agricultural Electronics Association was founded by the Equipment Manufacturers Institute in 1995 to bring together diverse interests in the field of electronics in agriculture. Membership has grown from 19 original members to over 100 companies, organisations, users, and university and government liaison members. Subdivisions within AEA include a User Council, Equipment Council, Hardware Council, and Software/Information Systems Council.

AEA identifies, develops, and facilitates appropriate action to increase compatibility and interchangeability of electronics and information systems used in agriculture. AEA has made significant strides in promoting standardisation, including addressing issues such as

— the interface between electronic equipment and specific connector, data, power and protocol requirements;

— compatibility of electronics and information systems with precision farming software;

— a standardised data "reader" interface between chemical labels and machines;

— environmental standards;

— livestock issues;

— development of a database standard;

— a common communication structure;

— standardised static and dynamic spatial data exchange formats; and

— data dictionary specifications in ISO 8211 format for yield (grain crops), soil fertility, and application crop plan characteristics.

Unbiased Evaluation

Unbiased, systematic, rigorous evaluations of the economic and environmental benefits and costs of precision agricultural methods are needed. An appropriate role for public agencies is independent, objective evaluation of precision agriculture technologies. Private technology development firms and input suppliers have a natural commercial interest in promoting precision agriculture. Individual producers may have insufficient incentives or resources to conduct evaluations or make the results known because all producers in a region can learn from those experiences at little cost, creating a "free rider" problem.

The benefits of having local information about the performance of precision agriculture technologies exceed those a single producer can gain, arguing for a public role. Moreover, producers may find it difficult to apply the experiences of a single farm to their own situations because they may not be able to make the appropriate adjustments for differences in conditions across sites. Site-specific factors can be so important in evaluating these technologies that the usual producer network will likely be inadequate for disseminating precision agriculture information.

Producers need unbiased assessments of precision agriculture's performance characteristics under various conditions. Public and private environmental organisations are also interested in unbiased evaluations of precision agriculture's environmental performance. Acceptance and support for precision agriculture depends on the extent to which potential efficiency gains and environmental benefits are actually achieved.

USDA is in a unique position to facilitate and coordinate evaluation and research activities among federal agencies. USDA and its affiliated SAES partners have the agronomic knowledge necessary to evaluate the effectiveness of specific precision agriculture technologies and systems. Where federal agencies outside agriculture have some basic technological components and expertise necessary to advance precision agriculture, collaboration in that evaluation should be encouraged.

Producers and other customers for precision agriculture technologies should be encouraged to search for multiple sources of information when deciding whether to adopt particular components of precision agriculture technology. Producer decision-making processes are complex, and multiple sources of information would help to shape and confirm decisions. Evaluation does not necessarily imply that producers are not capable of making sound decisions or that input suppliers are untruthful in their claims. Research in the area of decision theory indicates that sound judgments are particularly difficult for humans in situations where there is considerable variability, where there are time lags between actions and results, and where there are multiple and complex cause-and-effect relationships. These three characteristics seem to apply to precision agriculture and its adoption and evaluation.

Several factors will make it difficult for public agencies to carry out such evaluations in the area of precision agriculture. The technology is changing so rapidly that evaluations of specific technology components have a very short useful life. Producers will not be well-served if equipment or products embodying technologies which have been evaluated cannot be compared with newer, more sophisticated versions that have not been evaluated. Products that have been validated in the field before submission for evaluation, whether prototypes by industry or federal expertise, should be compared on the same basis as existing commercial products.

System evaluations are appropriate on technologies that are installed, maintained, and operated as specified by the manufacturer. Because the area is evolving so rapidly, technology developers may be reluctant to expose newly developed technology to public evaluation, risking loss of proprietary and trade secret information. Meaningful collaboration between private firms and public agencies, and between agencies, may not be forthcoming without considerable effort.

Over the long term, there is no substitute for carefully designed observation of economic and environmental results obtained by actual producers in real field conditions. For these experiments to be useful, side-by-side treatments and statistical control methods need to be used to distinguish precision agriculture technology's contributions from normal variation in resources, weather, and management. Given the systems nature of precision agriculture techniques and the importance of site-specific variability, on-farm experimentation performed in collaboration with producers will be necessary and desirable, compared with more traditional farms of plot-based research design. Research collaborators can mine a wealth of on-farm data and use regression analyses and other multivariate statistical methods to isolate the multiple sources of variations that influence economic and environmental outcomes of precision agriculture. These findings can provide invaluable guidance to producers on the expected benefits from adoption of precision agriculture technologies in their particular setting.

The accuracy and reliability of methods for collecting precision data need to be evaluated to ensure confidence in grid soil-sampling schemes, directed sampling, and yield monitor results. Similarly, the accuracy and reliability of methods for making precision applications of fertilisers, pesticides, irrigation water, and other inputs also need confirmation. Finally, producers need confirmation that new methods of interpreting precision input information to develop recommendations for management changes are accurate and do improve economic and environmental performance over whole-field management methods. Subfield fertiliser response relationships, economic pest control thresholds, and computer-based decision support systems derived from crop growth simulations need to be tested and evaluated under actual field conditions in a variety of circumstances.

Precision agriculture evaluation activities should be undertaken by both the public and private sectors. Organisations in both sectors should work together to avoid possible biases in evaluating the efficacy of the technologies. Because of the site-specific nature of the farm fields that precision agriculture is designed to address, evaluation cannot be generalised and must be couched in terms of the specific resource conditions to which it is applied. Evaluations should compare precision agriculture systems with conventional, uniform management systems, and against each other, recognising that precision agriculture enables changes beyond variable-rate application of inputs. Evaluation is part of an iterative continuous cycle of research, development, and deployment that is necessary for the technologies to evolve and improve.

Development of New Approaches

Precision agriculture requires new approaches to research that are designed explicitly to improve understanding of the complex interactions between multiple factors affecting crop growth and farm decision making. The most important research area of precision agriculture is development of theoretical and empirical knowledge to support improved crop models, farm management methods, and expert systems software. Much of the discussion of precision farming has revolved around measurement of variability through yield monitors, remote sensing, and digitised soil mapping. However, measurement means little if it does not result in better management. In this regard, precision agriculture is a systems approach to agricultural management, not dissimilar to the application of systems principles in other arenas since the 1950s, including environmental problems in the 1970s, IPM and sustainable agricultural systems in the 1980s, and watershed and ecosystem management initiatives in the 1990s.

Precision agriculture is fundamentally an information technology that focuses people on the complexities and interrelatedness of agroecosystems in a holistic way. If systematic understanding of the cropping system is not captured in models, however, it is unlikely that the volumes of precision data on subfield variation can be meaningfully processed to provide improved management decisions. If such models cannot be developed, much of the potential of precision agriculture will not be realised.

Reliable crop models are the foundation of any attempt to construct data-driven, computer-based decision support systems that can effectively use precision data to make precision management changes, yet few such models exist. Theoretical and empirical understanding of crop yield responses to variations in nutrients and soil quality remains primitive. For example, fertiliser recommendations are based on rules of thumb, such as "one and a quarter pounds of nitrogen per acre for every bushel of yield goal for corn," even though theory and evidence indicate that crop nutrient response is nonlinear and that yield will not respond to additions of some nutrients when others are limiting. Similarly, it is well-known in principle that crop productivity and crop nutrient response depend on elements of soil quality and tilth (such as structure, texture, organic matter, and water-holding capacity), yet there exists little quantitative modelling of these factors in crop production.

Pest management presents a similar picture; treatment thresholds are frequently based on rules of thumb because there are no reliable crop-pest ecosystem models. Even if rules of thumb for nutrient and pest management are based on significant experimental data, the resulting recommendations are developed on an aggregated basis that ignores the other factors that vary within fields. "[Precision agriculture] departs from current nitrogen fertiliser guidelines that were primarily developed on a regional scale. ... As a result, current nitrogen recommendations may have limited application to site-specific nitrogen management.". If modelled relationships can be developed that capture the effects of variation in factors that vary at the subfield level, better recommendations can be made.

Nutrients, pest management, and soil quality are obvious targets for public research because of their linkage to environmental quality. Nutrient pollution of both surface water and groundwater is a significant problem throughout the United States, and agriculture is a major contributor to nutrient pollution in many areas, such as the Great Lakes region, the Chesapeake Bay watershed, and the Mississippi drainage.

In many areas, reductions in nitrogen pollution resulting from improved nitrogen management could justify the use of variable-rate application and other precision farming methods even when reductions in fertiliser expenditures do not. Public incentives, such as regulations, cost-sharing, or incentive payments, may be needed to spur adoption in such areas, if it is not otherwise profitable. Models of crop-pest interactions are important for devising improved ecologically-based pest management strategies. More judicious use of pesticides could reduce environmental damage. Investment in soil quality is a central tenet of sustainable farming systems. Models elucidating the relative contributions of the components of soil quality to crop yield could help improve the design of sustainable farming systems.

Development of such crop models will involve basic research into the effect of subfield variation in availability of nutrients for crop growth, the effect of variation in soil quality on crop nutrient uptake and pest (insect, weed, and disease) prevalence, and crop-pest ecosystem interactions. It will also involve applied empirical research attempting to quantify crop yield responses to nutrients, soil quality, and pest prevalence across the growing season for important crops in different locations. It will require interdisciplinary study involving agronomists, entomologists, economists, plant pathologists, weed scientists, and ecologists. Developing such models will not be an easy undertaking, and efforts requiring coordination across many disciplines are essential.

The current research model for crop sciences employs carefully controlled research plot experiments with replicated block designs in which only the factor of interest (i.e., fertiliser rate) is allowed to vary. Because precision agriculture has already been identified as a technology of some promise, detailed small-plot studies and technology-adaptation experiments may not be necessary. Further, the system may need to evolve so that innovation and learning can exploit both traditional research plot experiments and information captured from actual field operations through precision agriculture.

Precision agriculture has the potential to collect many layers of data for entire fields and record detailed variation in many variables that affect crop growth. Thus, precision agriculture could change the research paradigm from station-based plot studies to farm-based studies by (a) using more complex experimental designs, such as incomplete block designs, row-column designs, nearest neighbour designs, and split plot designs; (b) specifically incorporating spatial variability in experimental designs; (c) supplementing mean-based analyses with comparisons of entire distributions; and (d) using statistical methods such as multiple regression. Under this paradigm, groups of producers would collect precision data; agronomic researchers would analyse the data with statistical methods to estimate how small changes in manageable factors affect crop yield in various resource and weather situations. The incentives for and obstacles to producer data sharing need to be fully explored and carefully understood if such cooperative on-farm research is to succeed.

Improved farm management methods are equally necessary. At present, farm management models are based on budgeting, which assumes fixed input-output ratios. Recommendations based on improved crop models will likely need to be derived from nonlinear optimisation, such as profit maximisation, because such crop models will likely be characterised by variable input-output ratios which may change throughout the growing season. Stochastic optimisation frameworks may be needed to take into account the random occurrence of rainfall. Dynamic optimisation may also be needed to take into account changes in recommendations as the growing season progresses.

It is not clear a priori that precision agriculture technologies will reduce environmental spillovers from agriculture. For example, if increased precision in application rates increases crop yields sufficiently, then producers may have an incentive to increase rather than decrease application rates. Thus, research explicitly aimed at elucidating the environmental effects of precision agriculture technologies should also be a priority.

Decision support system models force consideration of factors and interactions affecting the entire system. Therefore, work oriented toward developing such models requires institutional arrangements that cut across strict disciplinary lines. The relationships represented in such models are also inherently multivariate, lending themselves more toward on-farm research than plot-based studies. Reorienting research priorities toward model development and validation could thus alter institutional incentives for more holistic, farm-based, decision-oriented research efforts.

New Training Methods

In the twenty-first century agricultural professionals using information technologies will play an increasingly important role in crop production and natural resource management. It is imperative that educational institutions modify their curricula and teaching methods to educate and train students and professionals in the interdisciplinary approaches underlying precision agriculture.

For use of precision agriculture to become widespread, producers and prospective employees will need general computing skills and technical literacy. Specific skills needed by service specialists, such as high-tech equipment operators and GIS or GPS technicians, could be taught both through traditional four-year programs and vocational training. Consultants, system integrators, and others with an understanding of how to develop and apply precision agriculture will likely need postgraduate education.

Successful training of users of information technologies will require disciplinary depth (i.e., agronomy, agricultural engineering, and soils) and analytic skills (i.e., spatial analysis and crop modelling), that is best provided by long-term education with emphasis on interdisciplinary synthesis. The mind-set that is needed to ensure the beneficial use of precision agriculture should be fostered in educational institutions, particularly in elements of programs that provide an understanding of technologies in a broader context. While some institutions have developed undergraduate courses and extension education programs in precision agriculture technologies, a more systematic reappraisal of their programs is needed.

The content and form of agricultural education and extension are evolving under a number of pressures. Significant impetus for change comes from the declining public support for public education institutions in state and federal budgets. Increasing use of several kinds of distance education and on-the-farm research are examples of how the form of education and extension are evolving.

As part of the transition of parts of agriculture from agrarian to industrial processes, new domains such as precision agriculture are exerting a demand for new services. In this case, it is a demand for help in understanding whether and how to use the technologies and a demand for skilled graduates to design, implement, and operate the .systems. Some of these training and education gaps are being filled by the private sector. A significant challenge faced by public education institutions is to evolve a role that allows them to serve a broader social purpose, to guide the technology and turn out graduates who understand precision agriculture in a broader socioeconomic and environmental context, while still providing useful practical skills to graduates and useful information to users of the technology.

Using systematic, holistic, information-driven production management as an organising principle for agricultural education shifts the focus to encompass both the broader educational goals and the details of precision agriculture systems within a new framework. This synthesis could evolve from replacing narrow disciplinary frameworks with the concept of systems derived from crop modelling work.

Given the development of precision agricultural technologies outside traditional agricultural institutions, it is not clear who should provide scientific, technical, and managerial education needed for precision agriculture. In many cases, this expertise is provided by several university departments. Although it is unlikely that computer, geography, and engineering departments will begin teaching agronomy and other agricultural sciences necessary for precision agriculture, agricultural departments should seriously consider recruiting students with backgrounds in computer, geography, and engineering technologies that are driving precision agriculture.

It needs to be determined whether, in the short run, it is more efficient to teach basics of agricultural sciences to students already familiar with the technologies used in precision agriculture or to do the reverse. Designers of continuing education curricula will perhaps have an easier task. Material will have to be developed for both technologically sophisticated students needing more agricultural science background and agriculturists needing more training in the technologies. The wisest course may be to overcome institutional barriers on campus and develop creative syntheses between computer, geography, and engineering departments and agricultural science departments for courses that draw on the strengths of each.

Similar to the demand being placed on research in public institutions, extension agents, consultants, suppliers, and others who "retail" information to farmers will have corresponding requests to provide help with understanding and using the technologies. Professional organisations involved in precision agriculture, such as the Alliance on Agricultural Information Technology, have already called for "clearinghouses," unbiased sources of comparative information on alternative precision agriculture methods which could be set up and operated by extension services or others who work directly with producers.

A broader public purpose would be served if the clearinghouse function included unbiased comparisons of benefits, costs, and effects of precision agriculture adoption and use. Information service providers could help address specific problems that arise from precision agriculture and could play a role in promoting socially beneficial aspects, such as environmentally sound approaches. In the legal arena, templates, forms, and model contracts are needed to avoid conflicts associated with data ownership, intellectual property rights, and the protection of privacy.

The extension system would seem to be suited to such a public information role, but deficiencies in specialised scientific and technical training of current agents raises questions of capacity and capability. The rapid technological changes occurring in diverse areas of precision agriculture, including chemical pest management, environmental improvement, and farm electronics, present an enormous challenge to extension.

Moreover, the need to integrate these technologies within a systems view of agricultural management makes its leadership in precision agriculture information problematic.

High-speed Data Connectivity

High-speed data connectivity is needed in rural areas to support precision agriculture. Agricultural organisations and agencies should work collaboratively with public agencies and industries to ensure adequate rural connectivity. The communications infrastructure such as satellites, high-speed telecommunications services, and the Internet will be essential if precision agriculture is to develop to its full potential. Extensive adoption of precision agriculture will depend on access to a modern information infrastructure in rural areas, particularly telecommunication services such as the Internet. Such communications services will similarly be essential for taking full advantage of the data-generation capabilities of precision agriculture technologies described previously.

Private industry may provide little of this communications infrastructure. Infrastructure investments are characterised by substantial economies of scale because the cost of building the infrastructure is high, whereas the cost of providing service to additional customers once the infrastructure is in place tends to be low. Firms can recoup the needed investment in infrastructure by charging sufficiently high access fees and usage charges. Such charges will tend to inhibit use, however, so that use by some firms and individuals will remain below efficient levels.

Moreover, private industry may not find it profitable to provide infrastructure coverage to all of the areas of the country. In this case, benefits to society as a whole can exceed the investment costs. Rural areas may be particularly at risk because customer density is low. Development of the communications infrastructure may be an important stimulus for rural economic development.

In US, the Telecommunications Act of 1996, although not specifically addressing precision agriculture, regulates the competitive framework within which information technology services will be provided in rural areas. Within that framework, state telecommunications regulatory agencies, state and local governments and other rural institutions, such as university extension departments, will have an interest in encouraging rural access to advanced telecommunications. Access to skilled labour and up-to-date information about technical processes and market conditions have been important reasons for concentrating industries in certain locations in the past, leading to faster growth in urban than in rural areas. Modern communications are reducing the advantages of geographic concentration.

If competitors of the United States in global food and fiber markets adopt precision agriculture technology more readily than do domestic producers and are perceived to gain an economic advantage, public policies encouraging precision agriculture adoption may be developed to maintain global competitiveness. The President's Council on Sustainable Development found precision agriculture to be one of the environmental technologies that offer potential for domestic economic development through increases to domestic productivity and development of export markets in environmental technology.

Although the notion of universal access to telecommunications has been embodied in federal policy documents such as Principles for a National Information Infrastructure, it is unclear whether and how universal access will become a reality. Some experts suggest that the competition created by the Telecommunications Act of 1996 will be sufficient; others believe federal subsidies comparable to the rural electrification programs will be needed. In either case, access to information services will affect rural areas. To the extent that it takes place in different forms and at different rates of implementation, equity issues will arise. Advocacy groups are already concerned about disenfranchisement of information-poor socioeconomic groups. This potential exists in rural areas, particularly areas without activities such as precision agriculture providing an impetus for commercial enterprises to provide services.

Fast, reliable Internet access in remote rural areas will affect precision agriculture approaches that rely on regional aggregation and distribution of data or communication with experts, such as crop monitoring or interpretations of field data. The Alliance on Agricultural Information Technology identified rural bandwidth, GPS differential correction, and digital orthophotography as key information technology components of precision agriculture that currently restrict its adoption.

The components of an information infrastructure for precision agriculture include access to spatial technologies such as GPS and remotely sensed imagery, which will spill over into areas such as modernisation of land records and land use planning. In much of the rural United States, local government land information systems are antiquated and unable to provide timely, reliable information for private or government land-related decision making. In some areas, precision agriculture will be a useful force in improving these systems, in addition to providing some of the automated data and spatial technologies that will be required.

Intellectual Property Rights and Data Ownership

Precision agriculture will require clarification of intellectual property, data ownership, and data privacy rights. The extension service should play a leadership role in providing education on existing law pertaining to these issues. Intellectual property rights and data ownership are evolving areas of concern in terms of both information technologies and legislative and judicial activity. Precision agriculture raises some unique questions relative to data ownership because of the spatially extensive nature of the resources involved and because of its related information gathering systems, including remote sensing and third-party data providers.

Despite the continuing evolution of intellectual property law, legal precedents from the computer industry and general business practice provide guidelines applicable to precision agriculture. Legal complications need not constrain adoption of precision agriculture technologies if the legal forms from other industries (copyright, trade secrets, and patents) can be translated to precision agriculture, allowing producers to abandon handshake agreements and formalise their legal rights to their data.

Data ownership issues could affect the adoption and value of precision agriculture. A balance between protections for individual data ownership and benefits to multiple users must be found. Two scenarios could result:

— If ownership cannot be or is not protected, there may be a chilling effect on the willingness of individuals to provide field and farm data to aggregate databases, whether publicly or privately established.

— If ownership protections require producers to jealously guard their individual data, broadening the value of individual data collection through regional aggregation for area-wide crop management research or recommendations will be retarded and made prohibitively expensive.

Providers of information services, such as fertiliser dealers providing precision application services, generally recognise the land owner or farm manager as the *de jure* owner of data. This probably is supported by law, because the producer is buying the information services from the provider. Ownership issues may be further complicated by contractual arrangements between producers and providers, data acquisitions or exchanges by providers or third parties, and attempts to copyright data compilations. However, mere possession of the data by service providers, supported by the provider's intellectual investment in storing and analysing the data, may lead to appropriation of data rights, unless the operator's rights are specified with appropriate legal documents.

Although these concerns are novel for agriculture, many of the same issues have been faced by other industries. Some tools, such as copyright, cannot protect raw data but can protect the expression of ideas or concepts embodied in the data, such as a set of recommendations, a computer model, or a compilation of the data. Trade secret protection does not apply to raw data unless the producer can show that the data he collects, or pays to have collected for him, meet the specific criteria for trade secrets.

The data must derive independent actual or potential economic value from not being generally known to other persons who could profit by it, and the producer must show that reasonable efforts were made to maintain that secrecy, such as a nondisclosure agreement, a license agreement, or some other legal instrument that restricts access and disclosure to others. Once such legal forms deriving from other industries are adapted to the peculiarities of precision agriculture, many of the issues of data ownership will be resolved.

Because of the proprietary nature of computerised systems, the producer may get data in a format useable only by the provider, who may then exercise *de facto* control over the data. These are not legal ownership issues, but issues of technology and technological competence. The producer's recourse, in this example, may be the expensive one of paying for data conversion to the nonproprietary format so another provider can work with the data. This may be financially onerous but it presents no legal barriers to clear data ownership.

The value of individual field and farm data increases when it is collected across a region and integrated with data from other sources and other farms (see section on data assembly and aggregation, below). Aggregating data has implications for ownership as well. Some farm organisations have asserted a right to create such collections, presumably to provide better information services to members and to develop regional strategies.

Once the producer fails to assert ownership over precision data, compilations and abstractions from that data may be difficult to protect. If data are not handled as trade secrets, they become part of the public domain and proprietary rights are lost. Although raw data cannot be copyrighted, compilations of raw data that have been selected and arranged in creative ways can be protected through copyright. Vendors with regional databases can protect content to the extent allowed by merger doctrine. "When the expression of an idea is inseparable from the idea itself, the expression and the idea merge. ...". That is, copyright can protect a unique way of managing, analysing, or displaying the data that merges inseparably with the data and that cannot otherwise be copyrighted.

Intellectual property rights in precision agriculture products and software are protected in the same way that other computer hardware and software are protected. Copyright and patent laws apply to these creations just as for creative products of other industries. In general, the more basic the scientific finding underlying a new development, the less protection these traditional property rights instruments afford. Patents and other forms of intellectual property rights to knowledge were created to mitigate the problem of underinvestment in research and development. Patents confer a temporary monopoly (17 years) on the fruits of research, allowing patent owners to recover the costs of research.

Another body of law governs privacy issues associated with ownership and use of data by governments relative to individuals. Traditional privacy issues such as personal information on health and income are protected from disclosure by statute. For example, responses to the Census of Agriculture are protected from disclosure and can be published only in aggregate form. Land information in the public record is generally considered part of public domain (i.e., land ownership and real property taxes), including farm ownership and management information for federal conservation cross-compliance and many state programs.

Some forms of remote sensing may already cross the threshold into invasion of privacy (depending on purpose, access, etc.). This may depend on the degree of intrusiveness or "subjective expectation of privacy". Regulatory access is often a test of the Fourth Amendment (illegal search and seizure). The boundaries between public and private collection and use (open records laws) may make it difficult to determine what is subject to the Freedom of Information Act. A patchwork of laws, often poorly enforced, addresses restrictions on public agency use and disclosure of individual data.

In GIS, private sector operations, fearing loss of data or proprietary advantage, have been reluctant to participate in multipurpose land information systems where their data would be intermingled with other data in a public system (i.e., private utilities participating in local land records systems). By analogy, public-private cooperation in precision agriculture could be inhibited. Several groups advocated changes to current law to clarify intellectual property rights in databases.

Producers have expressed reservation that precision agriculture data may be used by government agencies for regulatory purposes. These new sources of information, however, will have the same privacy protections against use by government agencies as traditional sources of farm information, such as farm records, weigh bills, and other private documents.

Data Assembly and Aggregation

Data collected for use at the subfield and field levels have additional value for research, testing, evaluation, and marketing when assembled into regional databases. Mechanisms are needed to create these databases and make the data available for these additional uses including data collection and transfer standards; institutions for collecting, managing, or networking data; and policies to facilitate data sharing and access, while protecting proprietary interests and confidentiality.

As valuable as precision agriculture data may prove to be to individual landowners, much of the potential value of the huge amount of electronic data that could be collected by these technologies will not be realised unless the individual farm databases are consolidated into regional databases. These would not be averages or other statistical summaries of detailed data, but massive compilations of the detailed data itself, without information identifying individual farms from which the data are collected. Summaries might be made from these data for some purposes, but the detailed data needs to be accessible for analysis and modelling of relationships between inputs and outputs, including environmental outcomes; the data also needs to be sufficient to control for other sources of variation. Analysis and modelling of crop responses to a variety of soil, weather, pest, and other crop management factors can best be done when the most complete range of variation is present in the data.

An individual producer's database is limited to the trials and management actions that form the recent history on his or her farm, whereas a regional database would encompass a far greater variety of management responses to similar conditions. A regional database might include conditions that did not occur on an individual farm one year but could occur in the following years. Researchers, extension agents, input suppliers, commodity companies, and government officials as well as producers will be interested in such regional databases if they can be made available.

Despite the apparent benefits accruing from combining data into regional databases, it is by no means certain or inevitable that such data will be assembled. Information service providers that currently collect or are given precision agricultural data have proprietary interests in restricting access to their customer base, or in getting remuneration for access. In the absence of legal safeguards for data privacy, producers may be reluctant to share data on their operations in a freely accessible, voluntary regional database. Public agencies may not have the organisation, knowledge, or resources to develop regional data sharing cooperatives that could allow effective use of such data.

Two kinds of obstacles stand in the way of creating regional databases from farm and field microdata. The first, and more surmountable, are technical barriers such as computer capacity (whether centralised or distributed as a network) and transfer standards and protocols. Second are institutional barriers, such as lack of clear leadership roles in establishing the databases, legal issues of data ownership and privacy discussed above, and issues of compensation and access.

Widespread adoption of precision agriculture technologies will constitute a new way to practice agriculture at ever finer spatial and temporal resolutions, and to improve use of information for crop management at all spatial scales. These new capabilities offer the potential for a more economically and environmentally efficient agricultural sector. However, precision agriculture technology is new and largely unproven. Widespread adoption depends on economic gains outstripping the costs of the technology.

Exploiting the full potential of precision agriculture for environmental management will require fundamental shifts in public and private incentives for environmental management, and may require cost-sharing or other incentives for adoption. Lessons from the adoption of other agricultural and information technologies urge caution in anticipating the growth of precision agriculture use. Widespread adoption of precision agriculture methods will create changes in farm operations and in social institutions that can be anticipated and, where they are negative, mitigated.

Bibliography

Akridge, J. T., and L. D. Whipker., "Precision agriculture services dealership survey results", *Staff Paper* No. 97-10. Lafayette, Ind.: Center for Agricultural Business, Purdue University, 1996.

Altieri, M.A., *Agroecology: The Scientific Basis of Alternative Agriculture,* Westview Press. Boulder, Colorado, 1987.

Anderson, G.L. and C. Yang, "Multispectral videography and geographic information systems for site-specific farm management", 1996.

Auernhammer, H., M. Demmel, K. Muhr, J. Rottmeier, and K. Wild, *GPS for yield mapping on combines,* Comp. Elec. in Agric., 1994.

Barnes, E.M., T.R. Clarke, M.S. Moran and P.J. Pinter, Jr., Multispectral remote sensing and site specific agriculture: examples of current technology and future possibilities, *In Precision Agriculture,* 1996.

Birrell, S., and K. A. Sudduth, "Corn population sensor for precision farming", *ASAE Paper* No. 951334, St. Joseph, Mich.: American Society of Agricultural Engineers, 1995.

Carbone, G.J., S. Narumalani and M. King, "Application of remote sensing and GIS technologies with physiological crop models", Photogram. Eng. Remote Sens., 1996.

Chambers, R., A. Pacey, and L. A. Thrupp. (Eds.), *Farmer First: Farmer Innovation and Agricultural Research,* New York: Bootstrap Press, 1989.

Colburn, J., "Soil Chemical Sensor and Precision Agricultural Chemical Delivery System and Method", U.S. patent 1991.

——————., "Soil Constituent Sensor and Precision Agrichemical Delivery System and Method", U.S. Patent, Notice of Allowance, 1997.

Edens, T.C., C. Fridgen, and S.L. Battenfield, "Sustainable agriculture and integrated farming systems", *Conference Proceedings*, Michigan State Univ. Press, East Lansing, Michigan, 1985.

Grace, J.A., R.S. Pearson, and K.T. Gilmore, "The application of GIS technologies to prescription farming", *In Tech. Pap. of the ASPRS/ACSM Annual Convention*, New Orleans, LA, 15-18 Feb. 1993.

Lange, A.F., "Centimeter accuracy differential GPS for precision agricultural applications", *In Precision Agriculture, Proceedings of the 3rd International Conference*, Minneapolis, MN. ASA-CSSA-SSSA, Madison, WI, June 23-26, 1996.

——————., "Introduction to differential GPS for precision agriculture applications", *In Proc. Information Agriculture Conf.*, Urbana, IL. P&PI, Norcross, GA, July 30- Aug. 1, 1996.

Palmer, R., "Precise navigation, guidance and control services within the agricultural community", J. Navigation 42, 1989.

Index